Keeping Sheep

A PRACTICAL GUIDE

Foot trimming a ewe. This book covers all the sheep husbandry tasks a shepherd needs to know.

Keeping Sheep

A PRACTICAL GUIDE

Jack Cockburn

 THE CROWOOD PRESS

First published in 2023 by
The Crowood Press Ltd
Ramsbury, Marlborough
Wiltshire SN8 2HR

enquiries@crowood.com

www.crowood.com

British Library Cataloguing-in-Publication Data
A catalogue record for this book is available from the British Library.

ISBN 978 0 7198 4221 4

Photo credits
All pictures in this book supplied by:

Jack and Leah Cockburn
Peter Thewless
Pete Birkinshaw, Creative Commons (p.48, top left)
Ceredigion Museum Archives (p.8; p.153, right)
Dorset Sheep Breeders' Association (p.22, bottom; p.36)
Kendal Teeswaters (p.155, top)
Dave Lumb (p.50; p.85)
MacGregor Photography (p.22, top)
Hannah Watson (p.52; p.53; p.125; p.126, bottom; p.153, left)
Welsh Mountain Sheep Society/MacGregor Photography
(p.7; p.47; p.81, left; p.166)

Typeset by Simon and Sons

Cover design by Blue Sunflower Creative

Printed and bound in India by Replika Press Pvt. Ltd.

CONTENTS

FOREWORD

Britain is steeped in sheep-related history of which few people are aware in today's fast-moving world. But while it is right to understand the past, the role of sheep farming in the future is where we should focus our attention.

Today's global challenges of climate change, resource protection, nature recovery, and human health and well-being, are undeniable, and my belief is that sheep farming, with its glorious diversity, will play an important part in our future. Sheep farming can be almost the ultimate renewable technology, feeding on grasslands, reliant on few inputs other than sunshine, rain and soil, and producing highly nutritious lamb, mutton and sheep milk – and, of course, wool, which is once again growing in popularity as a natural, sustainable fibre.

Interest in keeping sheep is still very much alive and kicking, but keeping them well, in good health, and in a rich and vibrant environment, is essential not just for our reputation but to give us pride and satisfaction in continuing a practice that has been a big part of our past and will be a big part of our future. *Keeping Sheep* will be a useful resource for anyone starting a flock or wishing to improve their knowledge of sheep keeping.

Phil Stocker
Chief Executive, National Sheep Association

INTRODUCTION

The shepherding of sheep in Britain is as old as the hills, with the earliest evidence of their presence on these isles dating from the Neolithic period, 4000 to 3000 BC. The Romans were responsible for greatly increasing the number of sheep and the amount of pasture in Britain.

Welcome to the age-old pastime of sheep keeping. By keeping and breeding sheep today's shepherd is picking up a thread of culture and subsistence that leads back into the mists of time. Along with traditional farming, gardening and stewardship of the land, sheep farming can give the dedicated keeper a rootedness and purpose that is often lacking or difficult to find in the digital age.

SHEEP IN HISTORY AND CULTURE

For millennia, natural pastures have allowed the inhabitants of the island of Britain to produce sheep and harvest important products like food and wool for trade and export, but also for home consumption. Through the ages, sheep helped the people of Britain to survive the winters in the northern hemisphere by providing the raw ingredients for clothing, food and light (tallow for candles). Early civilisations in Britain realised that sheep can be produced on the hills and mountains, providing a supplementary winter food source to complement the crops that could only be grown on the fertile flatter lands.

Sheep in Britain have long occupied a key position in the national psyche, featuring prominently in art, religion and literature, from the Roman conquest through to medieval times and the romantic poets of the nineteenth century. The monks of the Middle Ages realised the economic benefits of organised sheep farming on the hills, and

Bringing sheep back to the farm from hill pasture, Rhydyfelin hill, Ceredigion, 1915.

sheep became the basis of their economic success.

SHEEP, LANDSCAPE AND WILDLIFE

The historic image of Britain in art, literature and song as a land of green rolling hills is woven from the ancient job of keeping sheep. Along with cattle, sheep are responsible for maintaining that green lustre for which the hills and valleys of Britain are renowned.

Without the nibbling of sheep, the landscape would take on a duller hue. Brambles, bracken and scrub would proliferate, rendering the land useless for food production and presenting a greater risk of wildfires during heatwaves. The landscape would also lose open grassland that many species of bird and insect have evolved to rely on. Sheep farming has shaped the evolution of wildlife in Britain over millennia.

Small-scale traditional sheep farming supports a rich diversity of natural habitats for wildlife. Your keeping of a small flock can have the added benefit of providing wildlife-rich grassland, field edge habitats of longer grasses, shrubs and seeding grasses, and the impetus for restoring old hedgerows. Before larger-scale, industrial-style farming predominated in the later twentieth century, our wildlife was rich and abundant.

Small-scale sheep farming and the conservation of rare breeds help to restore and rebuild what pesticides and intensive farming have destroyed while preserving the tools required for a self-sufficient agricultural system.

WHY SHEEP?

What has made this humble woolly herbivore take such a prominent position in people's minds, arguably more than any other animal?

The mild and moist maritime climate of Britain is perfect for growing grass, which is perfect for feeding sheep; this is the reason that the island has such a large national sheep flock (over 30 million, according to the National Sheep Association). Britain has

Sheep and meadowland. Native breeds are perfectly adapted to make best use of natural pastures and to maintain natural grasslands for carbon sequestration and wildlife.

considerable areas of hills, mountains and moorland, characterised by stony, steep or boggy landscape and acidic, thin soils, most of which cannot be cultivated for crops. Yet much of the area unsuitable for economic crop production will grow pasture all year round. These pastures sustain flocks of sheep and the sheep have enabled the people to sustain themselves.

THE PURPOSE OF THIS BOOK

Keeping Sheep details the sheep husbandry tasks that are required for successful home production of sheep. It gives a comprehensive and holistic view of sheep keeping, considering the changing seasons so that the reader can formulate an approach to sheep keeping that suits their own land, production objectives and lifestyle.

The book provides guidelines to when each stock task should be carried out, the reasons for it and ways to avoid health and welfare problems arising.

A wealth of 'how to' videos exist online, covering everything from shearing to foot trimming. Rather than seek to replicate this practical learning resource in print, I have aimed to provide useful background knowledge, to give the reader an insight into why each sheep husbandry task is done and what problems to look out for. On this basis, the reader will be able to sketch out a management plan for their own flock that covers the whole calendar year.

The importance of good land management in keeping a healthy flock is also covered. There are chapters dealing with the practical aspects of restoring and maintaining hedges and fences to keep the sheep flock secure and to benefit nature and the heritage of local landscapes. This leads on to how to manage the grassland effectively to keep the sheep healthy and well fed for as much of the calendar year as possible.

Ways of integrating regenerative agriculture, mob grazing and organic farming into flock and land management for economic and environmental benefit are also covered.

Chapters 6 and 7 look at animal welfare, selecting a suitable breed for your land and lifestyle, and marketing products from your flock. How to catch and handle sheep efficiently is discussed before the book moves on to dig more deeply into the practical elements of sheep keeping, for example dealing with common health problems such as lame sheep and internal and external parasites.

Looking after rams and the breeding cycle are covered as part of a guide to managing the flock through the different seasons, including winter feeding. Three important and informative chapters are dedicated to lambing time and caring for ewes and lambs.

Finally, the book rounds off with an in-depth look at shearing and sheep's wool, and even includes an introduction to training a sheepdog.

Conservation grazing and regenerative farming are important to reduce the environmental impact of sheep farming.

This book will provide the reader with a good base from which to sketch out a flock management plan that covers all the basic sheep husbandry requirements through a whole calendar year.

CHAPTER 1

FIELD BOUNDARIES

A healthy hedge dividing fields is good for shelter and fodder. Here you can see willow in the hedge, which can be a useful medicinal plant for sheep.

HEDGES – ANCIENT FIELD BOUNDARIES

The decision to start a sheep flock on your land can be the beginning of a countryside restoration and species conservation project. An old hedge is a thing of beauty and great utility and has a long, unsung history at its

roots. It serves many functions and looks very pretty in spring, summer and autumn, when it's festooned with berries and hips. Traditionally, hedges were relied upon to provide a stockproof barrier to keep sheep and cattle from straying and shelter from winter winds, and to protect crops from grazing animals.

BENEFITS OF HEDGES

Hedges provide wood for making useful things and logs for burning. A hedge boundary is also a source of wild foods and herbal medicines as well as a provider of shade on hot, sunny days and shelter from blizzards, torrential rain and chilling winds. On a cold and windy winter's day, the temperature can feel several degrees warmer behind the shelter of a thick hedge.

In winter, farm workers of old tended to the hedges by trimming them and 'laying' the hedge plants horizontally to block up holes and keep the hedge stockproof and healthy, with plenty of young thorny shoots. Cut twigs were stuck in the ground in gaps and would take root if there was sufficient moisture. It was time-consuming, labour-intensive but satisfying work. The process of cutting back renewed the hedge, the plants would live for longer and the new growth promoted flowering and fruiting, helping birds and pollinators like bees store up reserves for winter.

In summer, a hedge of mixed plant species is home to myriad insects for birds to feed on. Autumn brings seeds and fruits and places to hibernate. In spring, a thick-laid hedge provides ideal cover for farmland birds to nest in and shelters young lambs from cold winds. Look closely into old hedgerows and you can see gnarly old hedge tree trunks growing horizontally for sometimes 3m (10ft) or more, evidence of hedge laying carried out decades ago.

As farm workers moved to industrial areas in Britain or were lost in the wars of the twentieth century, hedges were no longer tended to in this age-old way. The postwar drive to increase food production and intensify and mechanise agriculture even led to the government paying farmers to rip out hedgerows. Between 1950 and 2007, over half a million kilometres of hedgerow were lost in Britain. Lack of management, development of land for building and roads and over-cutting the hedges back with tractor-mounted flails has also caused the decline of ancient hedges.

According to the Soil Association,[1] studies have shown that organic farms have on average 50 per cent more wildlife than conventional farms. Part of the reason for this is that they have looked after and preserved their hedgerows, regarding them as important wildlife habitat. Fortunately, some government-led schemes now offer grants to help farmers reinstate and restore old hedgerows.

Renewing Field Boundaries in Winter

Fencing is usually a winter job. When the nettles and annual weeds on the field edge have died back, it is easier to see the

A bumblebee collecting nectar from wild flowers (herb Robert) that grow along the farm's hedges.

'bare bones' of your field margin. Fences are usually installed tight to the hedge itself, as most farmers do not want to lose even an inch of grazing to a wild hedge advancing into the field; historically, this would have meant less field area for which to claim subsidy money. With regenerative and organic agriculture this is not such a concern, as hedges are valued for their shelter, beauty and historic and ecological value.

Hedges provide an alternative mineral-rich food source for browsing herbivores. In some areas of Britain, it is common to see the fence positioned on top of the hedge bank so sheep can graze both sides of the bank. This is not good for hedge plants, as the leaf buds are nibbled away. Animals grazing the bank cause soil erosion, leading over time to the loss of ancient field boundaries and important habitats for birds and invertebrates.

RESTORING HEDGEROWS

Hedges that are not tended to or well fenced begin to deteriorate as sheep and cattle walk through the gaps. Damaged hedgerows full of gaps become too open and exposed for birds to nest in.

[1] Bengtsson, J., Ahnström, J., & Weibull, A. C. (2005) 'The effects of organic agriculture on biodiversity and abundance: A meta-analysis' Journal of Applied Ecology, 42(2), 261-269.

If animals can walk over the hedge bank, they cause soil erosion and nibble off new green shoots of hedgerow plants.

Over time, animals walking through the hedge leads to large gaps forming and the loss of ancient field boundaries and shelter for animals.

A good, thick hedge growing low to the ground allows the build-up of a rich layer of leaf litter and composted organic matter under the hedge. This makes a wonderful home for hedgehogs, mice, frogs, toads, invertebrates, fungi, wild flowers and much more wildlife besides. An open hedge bottom, on the other hand, exposes the leaf litter to the wind and bare soil to the feet of cattle, sheep or rabbits, leaving it open to erosion and taking away an important living and feeding space for wild creatures.

Coppicing

Working on a hedge in winter in the traditional style, using hand tools, is a great way to immerse yourself in the gentle rhythm of the countryside and a wonderful opportunity to get to know the wildlife and landscape around you.

An old hedge line full of gaps with intermittent mature hedgerow trees and bushes can be restored easily over ten years and transformed into a young and vibrant field boundary once again.

Coppicing involves cutting off old hedge plants at the base in wintertime. After cutting, the plant will regrow with multiple stems. This age-old art allows harvesting of wood for different uses while stimulating natural renewal. In the first few years after coppicing, you will need to protect the young regrowth of the hedge from grazing animals. This means dismantling old wire fencing and replacing it with new fence materials or, if the wire is still good, renewing an old fence by replacing the posts holding it up with new ones.

Once the coppicing is complete, the hedge bottom is laid bare. Gaps of naked soil where no perennial hedge plants are growing can be planted up with new 'whips'. These are tree seedlings used to fill the gaps by planting 60cm (2ft) apart in a staggered double row.

The newly laid hedge should be fenced off from sheep. In the first few years you will see a strong regrowth of wild plants and flowers on the banks with the newly revitalised hedge growing through the middle.

After eight to ten years, you will have a fine, young, upright hedge that can now be laid. Basic hedge laying is achieved by trimming back the sides of the hedge and cutting out shorter or twisted gnarly pieces. The straightest upright stems are retained and laid by cutting through the stem at the base but leaving one third still attached to the root base. A hedge is usually laid in an uphill direction to help the sap rise to the ends of the stems in spring and promote new growth. Once laid down across each other, the stems are woven in and out of upright sticks that are banged into the earth at intervals of between 60 and 120cm (2-4ft). Small cuts can be

This hedge was coppiced and left to regrow for eight years. The new growth was then laid down by cutting the small trunks two-thirds of the way through and laying them on top of each other. Today it is a thick and vigorous hedge again, a good windbreak and a great habitat for wildlife.

made into the laid stems on the skyward side to promote new shoots of growth.

FARM FENCES

DIY fencing installed by new smallholders is often quite easy to spot by grizzled old farmers, who smile to themselves as they go bumping past in their tractor. The wire netting sags and bows between posts that stand too tall and at jaunty angles. The fence might even offer a proper belly laugh if the wire netting is fixed upside down, an easy mistake to make. And the incomers from town won't have used any barbed wire on top because of the bad karma associated with it, meaning cattle and horses can easily push it down.

Unfortunately, a saggy fence is not a secure one. Fencing materials are expensive, but if you do a good job, the fence will keep your animals where you want them for a decade or more. Poor fencing is a source of sheep-related stress to the shepherd. Sheep will rarely escape when you have time to round them up. Instead, it is usual for them

Tree guards are used to protect the saplings from the nibbling of voles and rabbits in their first few years of growth. Voles in particular can be an unseen menace to newly planted tree whips.

to choose a moment designed to cause maximum angst, such as when you're just leaving home to attend a wedding or an important job interview.

The Advantages of Well-Fenced Fields

'You can't farm without fences' goes the old saying, and it is largely true. 'Good fences make good neighbours' is probably the second most memorable nugget of rural wisdom I have heard many times. Farming without fences is more akin to ranching, and not a way to efficiently produce quality livestock.

Besides the obvious function of preventing your animals from straying and wandering into high-value crops such as vegetables and cereals (or your beloved garden flowers), fences are essential for good grazing management.

Fencing End to End

A line of fence is only as secure as the 'strainer' posts at each end of a length of wire. Take time to dig in solid end posts (13cm/5in in diameter or more) so that 60-90cm (2-3ft) of post is in the ground with 120-150cm (4-5ft) standing tall. Old telegraph poles cut up with a chainsaw are ideal for this, as they are solid and well creosoted. Dig in a flat-bottomed post using a metal bar to loosen the soil and use a shovel or post-hole digger tool to remove the loose soil and stones. Make the diameter of the hole as small as is possible so the post fits snugly inside. Measure the depth of the hole as you go; when it is the required depth, slide the post in and backfill with fine, stone-free earth, tamping it down with the end of the bar as you go.

Brace your first post in the direction you wish the fence to run. Nail a strand of barbed wire to the bottom of your strainer post (most livestock farmers put a strand at the top and bottom of fences). Unroll the barbed wire and use it to mark the finished fence line. Go as far as you can in a straight line up to 50m (164ft) from your starting point.

Once the second strainer post is in the ground, you can tension the bottom wire between the first and second post. This will give you a nice, straight guideline along which you can bang in the smaller intermediate posts every 3m (10ft). Barbed wire along the bottom of the fence stops the sheep from pushing the netting up. If you decide not to use it, a long piece of string or rope to temporarily mark your fence line can be used instead so that you get your posts in a line. You can buy a specialist hand tool designed for tensioning a single strand of wire. The tool grips the wire and is used as a lever to tension the wire around the end post. A staple is then driven into the post

If the fence line requires a change of direction or meets an obstacle, this will be the location for your next strainer post.

to secure the length of wire in its tensioned state. If working alone, three hands and the dexterity of an octopus is required for this manoeuvre, although you can often get away with using your knee to hold the strainer tool in place.

A specialist fencing contractor can bang strainer posts in with a large post knocker mounted on a tractor or digger. A strand of barbed or plain wire will be stretched between two strainer posts. Then smaller intermediate posts will be knocked in at intervals of about three paces. A tractor or digger is also useful for tensioning the stock netting. The netting is clamped with a special metal clamp designed for stock netting. One end of the netting is stapled to a post and the other end is clamped and attached to the tractor using a chain and ratchet. The wire is then tensioned and secured to the end strainer post with staples. This can be dangerous if too much tension is put on the wire, so be careful.

Stock Netting

This is the wire with square holes in it that you find along hedgerows wherever there are farm animals. In livestock-raising areas of

Tools for putting up agricultural livestock fences with barbed wire and stock netting (left to right): roll of stock netting, wire netting clamp tool, bucket of fence staples, 13cm (5in) nails, handsaw, chainsaw (PPE should be worn), post knocker tool, spade; (bottom) roll of barbed wire, length of chain and fence puller, fencing pliers, claw hammers, leather gloves.

the UK you will see it everywhere. It can be bought in neat rolls of 25m, 50m or 100m from agricultural merchants. A roll of wire is very heavy, so, unless you have long straight sections of fence to do, it's more practical to buy the shorter-length rolls. The wire is galvanised so it does not rust and it will outlive the wooden posts used to hold it four times over. Often an old saggy fence line can be rehabilitated by knocking in some new posts along its length to support the wire. Renew the strainer posts at each end of an old line of fence wire and it can be pulled tight again.

From personal experience, I can vouch that a mistake the novice fencer will often make is to staple the wire to the posts upside down. If you look closely, you will see that the square holes in the wire are smaller on one side. The smaller holes are supposed to be at the bottom of the fence and are designed to stop young lambs wandering through.

Fence Posts

This subject is a source of much discussion and angst for farmers. The natural life of softwood fence posts is disappointingly short. Most are made from treated softwoods and the portion in the ground rots away after five to seven years. This is the reason that you see so many farmers' fences falling over willy-nilly. It is worth spending more on longer-life posts if you can because putting up fences is costly and really hard work. Some posts will have a fifteen or twenty-five-year guarantee.

Poor-quality softwood posts will rot after four or five years. It is worth investing in longer-lasting posts for new farm fences.

In olden times, split oak posts were used, and these can still be seen in hedgerows, a hundred years later, while their softwood chemically treated cousins of less than ten years are rotting into the soil.

Post Banging

Without a tractor-mounted post knocker machine, banging the posts into the ground is the most physically demanding part of fencing. A long metal bar with a point on the end is useful for this job. The bar is speared into the ground on the spot where the post is intended to go, making a guide hole and testing the ground to see if there are rocks or tree roots in the way.

Once a hole has been made with the bar, the pointed end of the post can be slotted in and the post will stand up ready for you to hit it with one of your heavy implements. A post knocker tool consisting metal tube, 13–15cm (5–6in) in diameter, with handles welded on the sides, is better for the job than a large mallet. It is much easier to knock the posts in straight with and does not split the wood on the top of the post like a mallet or sledgehammer will. Be warned, however – when lifting the post knocker ready for another enthusiastic downward blow, be careful not to catch the edge of the metal tube on the top of the post. Many people have clonked themselves on the head doing this, as the weight of the heavy tool is diverted towards the head of the off-balance operative.

Tensioning and Stapling the Fence Wire

Once you have your line of posts in the ground, it is time to staple the stock netting to the posts using a claw hammer and large metal fencing staples. Check the netting is oriented with the smaller holes at the bottom then securely staple one end of the wire to your first large strainer post. Unroll the wire as far as the next strainer post. Pull in as much slack as possible by hand then attach your fence clamp device or clamp the wire using bolts between two 90cm-long (3ft) pieces of wood. The netting is clamped so that when it is pulled, all the strands of wire in the netting are pulled equally tight. Attach a chain to the wire clamp and secure it to your tractor, 4x4 car or digger arm. The vehicle point to which you attach the chain must be positioned in line with the line of posts.

Now use your fence-puller tool to tension the wire and draw the stock netting tight over the posts. Staple the netting securely to all the intermediate posts in between and double or triple staple the wire to the strainer post. Release the tension and remove the clamp from the netting. You should now have a beautiful straight and tensioned line of wire fencing that will keep your sheep exactly where you want them.

Fencing Contractors

Apart from a tractor-mounted post knocker, the tools required for fencing work are not too expensive. Consider, however, if you will have the time and physical capability to invest in the job. Fencing is an intensive workout of heavy lifting, squatting, pulling, bending, fetching and carrying and a great deal of walking up and down the stretch of hedge line that you are working on. It's a job that requires perseverance and determination.

You may quite reasonably decide your skills are best employed on other tasks and hire a fencing contractor. A contractor will usually charge by the metre of fence to supply the materials and put up the fence. It's worth discussing the quality of the posts with them and buying longer-life posts if you can afford them.

CHAPTER 2

MANAGING GRAZING AND GRASSLAND

A Llanwenog ewe and her three lambs in mid-west Wales.

The graziers of today can reap the benefits their predecessors have bequeathed, and use the native breeds of sheep that are perfectly adapted to make best use of the natural pastures. The modern shepherd can pick up the sheep tradition and produce climate-friendly food and wool of the highest quality. All of this requires good management of the pastures.

A well-managed pasture means well-fed, healthy livestock. The grazing and growing of good pasture should be a key objective for anybody keeping ruminant livestock. This chapter covers the main points of managing grassland so that you can get the best out of your sheep and the land you are looking after. This is the cornerstone of a productive, healthy sheep flock. Grass is the lowest-cost, most economical feed for sheep. The way the grass is managed is also the single most important factor in managing the health of the flock.

STOCKING RATES

If your pastures are well managed and not too heavily stocked with animals, you should be able to keep sheep all year round and require very little bought-in feed. Some hay to feed during harsh winter weather and perhaps some concentrate for prolific ewes to eat pre-lambing may be needed, but the bulk of the sheep's diet will be homegrown.

LIVESTOCK UNITS

Livestock units (LSU) is a standard measure of the grazing impact of different types of ruminant livestock, usually expressed as livestock units per hectare or LSU/H. It's a system that can be used to compare the stocking rates on different kinds of farms and in differing types of production system, for example intensive versus extensive (organic, natural, non-intensive), or high-input systems versus low-input systems (where inputs are feed, fertiliser, fuel, medicines, seeds, pesticides and so on).

1 dairy cow = 1 LSU
1 beef cow without calf = 0.80 LSU
1 growing beef animal (12–24 months) = 0.60 LSU
1 breeding ewe with lamb (60+kg/132+lb) = 0.15 LSU
1 sheep = 0.08 LSU
1 horse = 0.80 LSU

One hectare is 10,000 square metres or 2.47 acres. One acre is roughly the size of a football pitch.

There is no easy answer to the question of how many sheep can be kept per hectare.

Potential stocking rate will vary from region to region and even field to field on the same farm. For example, a sheltered level field in Herefordshire could produce five times more grass and of better quality than a field at the base of a Lakeland fell on thin, acidic soil. This is partly the reason that breeds adapted to local conditions have been developed and why so many distinct sheep breeds can be found in Britain.

Land is graded from 1 to 5 in the UK according to quality. For more information about how land is classified, search online for 'agricultural land classification UK'.

As a rough guide, you might expect to carry six to ten sheep per hectare (two to four

The number of grazing animals a piece of land can support will vary according to soil type, drainage, the age of the sward (sown pasture), the grass species present, the amount of clover present, altitude, aspect (north- or south-facing) and the longitude and latitude of the farm.

Scottish Blackface ewes. The regional contrasts in temperature and precipitation can be marked in the UK. While people in Sussex might be shearing the downland sheep and weaning lambs in May, shepherds in Scotland can be braving snow showers while they check on their Scottish Blackface ewes lambing on the hills.

A Dorset Horn ewe, a breed developed in the rolling green hills and lush pastures of Dorset, southwest England.

sheep per acre). This is a modest stocking rate to begin with. It takes time to get to know a piece of land and work out how many grazing animals it can support through the calendar year.

Mature ewes of around 50–60kg (110–132lb) will need approximately 0.6ha (1.5 acres) of grazing each in the winter. In the summer, 0.2ha (0.5 acres) each of grazing should be ample. In practice these figures

can be altered by feeding hay and using rotational grazing. Neither the sheep nor the pasture will thrive if the sheep are left on the same patch of ground all year, unless there is a very low stocking rate. If the pasture is short and turning to mud in winter, the sheep should either be moved to another field to give the ground a rest or they should be housed for eight to twelve of the coldest, wettest weeks so the ground can recover.

Case Study: Overwintering Lambs Without Feeding Hay or Concentrate

In early November, ninety-six seven-month-old hoggets were put on a 7ha (17-acre) south-facing block of fields in west Wales where there was a considerable amount of deferred grazing available (leftover summer grass). This equated to 13.7 sheep per hectare (5.6 sheep per acre). By the end of January, after twelve weeks, the sheep had to be moved as they were losing weight and had eaten all the nutritious parts of the pasture. Because of low light levels and temperatures, the grass was not replenishing itself daily and had drawn much of the nutrients back to the roots. To kick-start the hoggets' growth, forty of the males were moved to new pasture that had been sown the previous May with ryegrass and clover. The aim was to fatten them ready for sale during the Easter lamb trade and get a good price for organic grass-finished lamb (a hogget is a lamb between weaning and first shearing).

The 4ha (10 acres) of new pasture had not been grazed since the end of August so there was a considerable amount of grass cover available (stocked at 3.6 sheep per acre). The average weight of the hoggets in January was 34kg (75lb). They were sold in April at an average live weight of 40kg (88lb), after being fed exclusively on grass with a mineral bolus given to each animal. The 4ha (10-acre) block was then mowed for silage the following May. The sheep grazing had helped to thicken up the sward, encouraging the grass to produce more shoots.

The winter was relatively mild, with little frost and no snow, so hay was not required. This was a cheap, less labour-intensive way of fattening lambs to sell when the price is highest – the alternative would have been to house them and feed daily with hay and concentrates. This system of grass-finishing lambs is dependent on having a large area of winter grazing. If the hoggets had been left on the same block of permanent pasture all winter, they would not have grown until the grass growth accelerated in mid-April; they would then have been too old to sell as 'lamb' so would have lost considerable value.

FIELD MANAGEMENT

Well-fenced smaller fields give the farmer or smallholder the ability to manage the supply of grazed grass through all seasons. This gives the shepherd the opportunity to feed their sheep more efficiently and to manage the grassland for grazing, hay production and for wildlife if desired.

A supply of clean, fresh water in the field is essential for all sheep, especially for ewes producing milk for their lambs.

Electric Fencing

Fencing is a large capital cost for any farm or smallholding. If larger fields need dividing into smaller paddocks to manage the grazing, electric fencing can be a viable option. The equipment is lightweight, portable and cost-effective, though it does need careful setting up and maintaining to get the best out of it.

Electric fence manufacturers can advise on suitable electric fencing equipment for sheep and setting up. Once installed, the fencing must be checked every day alongside

Electric fencing, such as the two white polywires used in the picture, can be used by the shepherd to set up a system of rotational grazing where permanent fences are not available. This can help in managing the supply of grass and protect the health of the sheep by allowing them fresh pasture.

the usual sheep welfare checks. Daily checks should include:

Voltage An electric fencing energiser unit can be powered by mains electricity (if close to a supply) or by battery. A simple volt metre can be purchased that allows you to test the correct voltage is travelling through the fence wire.

Batteries A common reason for electric fencing not functioning properly and not keeping the livestock where you want them is the battery being low on power. In the winter, or in bad weather, it is necessary to carry the battery to the mains and charge it. A spare battery and battery charger is a necessary investment.

Entanglement Sheep are much more likely to become entangled in the fence wire if the voltage of the fence is low. If the sheep jump through the wire, they can easily become trapped. They can cause themselves serious injury by panicking and wrapping themselves in a poorly installed or maintained electric fence. It is worth investing in proper reels to hold the fence wire for when it's being moved or stored – it's all too easy to end up in a terrible knotted mess when attempting to cut corners in the setting up or moving of electric fences.

BALANCING THE SUPPLY OF GRASS OVER THE YEAR

It is difficult for the shepherd to balance the supply of grass with the number of sheep kept through the seasons, as most of the grass growth is between May and and September. On self-sufficient farms, the summer surplus of grass is turned into 'conserved forage' in the form of hay, haylage or silage, which is then kept for winter feed.

On smallholdings it may not be possible to do this because of limits to machinery available, number of acres and storage.

In the summer, a solar panel can be used to power the electric fence energiser and save work recharging batteries.

Large round bales of home-produced hay stored in the farm's barn. This is an efficient way to harvest and store surplus summer grass but the farmer must have a tractor with fore-end loader, amongst other equipment.

On many smallholdings the land is used for grazing all year round and hay must be bought from elsewhere. In this case it is still important to try to rest the pasture from grazing for some of the year if possible.

IMPROVING PRODUCTIVITY

Rotational Grazing

Sheep require securely fenced fields, particularly around the boundary. Ideally the grazing area will be split into separate fields or paddocks with a water supply in each or a mobile water supply provided as the sheep move. Grazing can then be managed rotationally, moving the sheep to fresh pasture every week or two. This helps to control internal parasites and will produce more forage from a given area than 'set stocking' (letting sheep have the run of the whole area all the time). Aim to keep grass length between 4 and 8cm (1.5-3in) for maximum production. Rotational

grazing benefits sheep health overall, as it regularly provides them with fresh forage to eat. A rotational grazing system also gives the pasture a rest from grazing, allowing it to renew itself and send reserves to the roots.

Raising Soil PH

Simple soil testing can reveal the pH (acidity/alkalinity) of the soil underlying your pasture. Home test kits can be bought online or soil samples can be sent away to a lab for a more comprehensive analysis of potassium, sulphur, phosphorus and organic matter content (agricultural merchants can advise on soil testing). The optimum soil pH for grass growth is slightly acidic, at 6-6.8.

Soils in Britain become more acidic over time due to the acidity of rainwater (unless the land is in an area of predominantly limestone or chalk bedrock). Soils in higher-rainfall areas, such as the western and northern parts of Britain and the uplands, suffer from greater acidification. Moss and a mat of dead and

A rotational grazing system produces good, fresh grass for your livestock.

Get to know your soil by digging test holes in different parts of a field. A healthy soil will have plenty of earthworms and a good spread of plant roots into the lower depths of the soil. A healthy soil is like a sponge, with a crumbly texture and plenty of air pockets that allow plant roots to breathe and aid in the storage and filtering of water.

The humble white clover, an extremely important plant in British organic agriculture. Nitrogen is harvested from the air by clover via a clever symbiotic relationship between bacteria and the plant's root nodules.

HARMFUL EFFECTS OF ACIDIFICATION

In acidic soils there is less biological activity from soil micro-organisms, resulting in slow rates of decomposition of organic matter. Low biological activity means nutrients are not made available to living plant roots thus slowing their growth and consequently producing less vegetation for sheep to graze on.

decaying vegetation on the surface of the soil are indicators of a low soil pH.

Limestone

Since ancient times, agricultural soils have been improved by spreading burnt limestone on the land. Boatloads of limestone were brought to communities along western and northern coasts. The limestone was 'burnt' or heated in kilns on the shoreside then crushed to a powder and carted away by horses or oxen to be spread onto agricultural fields. To go to so much trouble and expense, the people must have observed that lime had a hugely positive effect on crop yields and plant and animal health.

Today, 'prilled' lime (little pellets) can be bought in bags and spread with a fertiliser spinner on a tractor or quadbike or spread by hand over smaller areas. Ask for advice on spreading rates based upon your soil pH. Larger areas can be covered with ground or crushed limestone by an agricultural contractor with a tractor and spreader – ask in your area for contacts. Crushed limestone is slower release and longer acting than the prilled calcium fertiliser balls.

Soils with a pH of 5 or less will need more than one application of lime. It is best not to apply lime too heavily in one dressing but to dress the soil over a two-year period. Before adding lime to your fields, check that they are not designated as conservation grade (SSSI) acid grassland or have rare acid-loving meadow plants growing in them.

Encourage Clover

Raising the soil pH to 6 or above will encourage clover to thrive in your pasture. Clover can be introduced by spreading seed on bare patches of soil, but it will also appear naturally over time. Clover increases the fertility of the grassland as it acts as a natural fertiliser by fixing atmospheric nitrogen into the soil. It is also higher in protein than grasses and so is a valuable addition to a sheep's diet, especially for fattening lambs or feeding pregnant ewes. If clover is allowed to flower in the summer, it provides a valuable source of nectar to bees and other pollinators.

In areas where the soil is naturally acidic, white clover can be encouraged by balancing the soil pH with lime. Clover provides extra protein to the sheep's diet and also feeds the soil with nitrogen captured from the atmosphere.

Introducing White Clover

Correct the soil pH of your field before overseeding with clover. A pasture can be scratched with a rake or an agricultural harrow and oversown with clover seed in early summer. It is best to graze the pasture heavily beforehand and then allow the feet of livestock or a heavy roller to press the seed into the soil after sowing. If possible, do not graze for eight weeks after sowing to let the young clover plants establish.

Sheep grazing naturally increases the amount of clover in the pasture over time. This has long been known as the 'golden hoof' effect.

Mixed Grazing

There are benefits to mixed grazing with different species of ruminant animals either at the same time on a given field or consecutively. Mixed grazing can cut the amount of pasture wastage. Cows are less selective grazers than sheep, taking larger mouthfuls of grass as they wrap their long tongues around it. Sheep are picky grazers, seeking quality over quantity, whereas cattle operate a quantity-first strategy because larger ruminants require greater volumes of fibre to convert into energy.

Cattle grazing can be particularly beneficial to biodiversity in the uplands, where rough grasses such as *Molinia* and *Nardus* dominate.

Sheep will graze closer to the cows' dung pats than the cows will, which increases the amount of pasture that is utilised.

Mixed grazing has been shown to increase growth rates of both cattle and sheep and it plays a role in controlling internal parasites (*see* Chapter 16).

Cows are good at munching rougher areas of pasture that sheep will not eat. The munching of cows on tussocks and trampling with their feet lead to an increase in the diversity of plants in the sward.

The overall production of grass from a given acre is increased by mixed-species grazing. Mixed grazing has been shown to increase growth rates of both cattle and sheep, and it plays a role in controlling internal parasites (*see* Chapter 16).

Artificial Fertilisers – Nitrogen, Potassium, Phosphorus

Fertiliser allows the farmer or smallholder to keep more sheep per acre of land. Artificial fertilisers containing high levels of plant-available nitrogen are only necessary on grassland, however, if the land is being asked to keep more grazing animals than it would naturally be able to support with good husbandry techniques and natural fertilisers alone.

There are significant drawbacks to chemical fertilisers. Their production uses huge amounts of fossil fuels. They disrupt the balance of soil micro-organisms and fungi and, during wet weather, fertiliser run-off or leaching through the soil pollutes watercourses and ground water.

Manures

Farmyard manure is a mix of composted animal dung and straw (sometimes woodchip) that has been used for animal bedding. When composted and applied to the land, farmyard manure is a wonderful fertiliser, increasing the growth and health of plants. Composted manure is a natural soil conditioner: it stimulates biological activity, encourages earthworms and improves soil structure by adding organic matter. When a field is cropped for hay and the grass baled and taken off the field, nutrients are removed from the soil. Farmyard manure is used by farmers to replace these nutrients and keep the soil fertile and productive.

Mechanical Operations

If the ground is not too wet and the soil is dry enough to support machinery, improvements can be made to grassland by using machinery at certain times of the year. Seasonal grassland machinery work includes the tasks below.

Spring

- Harrowing with a spring-tine harrow or chain harrow to flatten molehills, rake out moss and dead plant material. If harrowing is done vigorously, it can disturb the top layer of soil and prepare it for overseeding the pasture with grass, clover and herbs
- Ploughing and power harrowing fields which are to be reseeded with grass or used to grow an alternative crop. Ploughing has a negative effect on soil structure and releases soil carbon but is an effective way to control weeds before re-sowing the ground with a new crop
- Rolling hay and silage fields with a heavy flat roller to press in stones and flatten bumps in the soil; this helps to prevent stones from damaging the mower later in the season and also stops soil contaminating silage bales

Midsummer

- Mowing for hay, tedding grass and baling
- Topping fields that are not going to be baled and leaving the grass to rot down
- Topping grazing fields to cut docks, thistles, rushes and other weed plants that grazing animals will not touch

Late Summer

- Spreading farmyard manure on fields that have yielded a hay crop. This should not be done if the field is needed for autumn grazing

Early Autumn

- Harrowing
- Cropping (for silage)
- Rolling
- Overseeding
- Muck spreading
- Liming (aim for soil pH 6)
- Aerating (mole ploughing, draining or subsoiling)
- Ploughing and re-seeding is quite a drastic measure but if done well it can produce excellent results and greatly boost the productivity of a given area. You may wish to use agricultural contractors to carry out machinery operations. *Farmers Weekly* publishes an annual guide to contractors' rates so you can get an idea of costs, although rates will be higher for smaller acreages

Avoid Soil Compaction

It's very important to only allow machines onto the land when soil conditions are firm and dry enough. Deep and lasting damage can be done to soil structure when vehicles are driven on grassland in wet conditions. Soil compaction caused by machinery goes deep through the soil profile and a soil can take decades to recover from heavy wheel marks. The important air pockets in the soil collapse under the weight of the wheels, meaning plant roots cannot 'breathe', and the resulting poor plant growth cannot easily be repaired.

Overseeding

If weather conditions permit, spring and late summer can be good times for overseeding existing pasture by harrowing, broadcasting grass and clover seed and rolling with a heavy roller. Ideally, the grass should be grazed down short beforehand and the surface area scratched up, so that 25 per cent of it is showing soil. The field can be grazed for the first two weeks after overseeding. The grazing animals press the seed into the soil with their feet. Livestock should be excluded from 2–10 weeks after seeding to allow the seedlings to establish.

PERMANENT PASTURE

Permanent pasture is defined as grassland that has not been ploughed for five years or more. In the twentieth century, hundreds of thousands of acres of permanent pasture were lost or degraded due to ploughing, intensive farming and use of herbicides and artificial fertilisers.

BENEFITS OF PERMANENT PASTURE

With good management, permanent pasture has four main benefits:

- It acts as a carbon sink. The soil is not disturbed and builds up stores of carbon over time.
- The increased carbon helps build structure and improves the 'sponginess' of the soil, allowing permanent pasture to store water during wet weather and reduce floodwater run-off from the land.
- Permanent pasture contains a more diverse range of herbs and grasses than fields sown with agriculturally improved grasses, and this diversity can benefit animal health.
- Plant diversity in permanent pasture is beneficial to wildlife such as birds and pollinating insects, especially when the plants are allowed to flower and set seed before hay making.

CHAPTER 3

REGENERATIVE AGRICULTURE AND ORGANIC SHEEP FARMING

Lambs grazing clover – an environmentally friendly way to produce meat while adding to the fertility of the soil.

Regenerative farming, organic farming, nature-friendly farming, biodynamic farming, agro-ecological farming – these are all approaches to land management that overlap considerably in their aims. The chief aim of all of them, however, is to produce food for human consumption while working in harmony with nature. In the marketing of foodstuffs, many claims are made about the environmental credentials of the food and production system, but organic farming is the only ecology-friendly food production system that has standards of production enshrined in law and an established inspection and auditing system.

Organic certification is a guarantee to the customer that food has been produced without pesticides and artificial fertilisers and with higher animal welfare standards.

In principle, the agricultural and environmental questions that the various ecological approaches to farming are all grappling with are similar:

- Can the farm system on a given acreage produce food from the soil indefinitely

Organic products must carry a logo from an organic certifying body to confirm the production system has been inspected and audited and the farm is complying with organic standards.

(sustainably) without external inputs (such as fertiliser and chemicals)?
- Is the farm system self-supporting (does it rely on heavy use of external inputs like fossil fuels, fertilisers, imported feeds)?
- Will your farm system increase the diversity of plants and wild creatures on the land and build soil health and fertility over the longer term?
- Can all the above fit into an economically viable farm business?

Big questions and plenty to think about!

REGENERATIVE FARMING

It is worth considering how your flock of sheep could fit into a regenerative approach to managing the land that you are farming.

The term 'regenerative agriculture' refers primarily to carrying out practices on the farm that rebuild soil health and soil carbon. These include restoring wildlife habitats such as hedgerows and hay meadows that have been lost or degraded.

Much of the regenerative approach uses methods from traditional mixed farming, resurrecting practices from a time when farmers had to rely on natural methods of increasing the fertility of the land and growing crops, as artificial fertilisers and animal medicines weren't available.

Rebuilding soil health and restoring wildlife habitat is achieved by moving away from the conventional farming practices that have been proven to damage the environment. A regenerative approach to agriculture would include:

- Dropping the use of herbicides, such as glyphosate, because of its negative effect on soil microbiology and plant diversity on the farm
- Ceasing to use artificial nitrogen fertilisers because of their negative effect on soil carbon and soil structure, and because of the heavy fossil fuel cost of their manufacture
- Stopping or reducing ploughing and cultivating soils because of the mechanical damage done to soil structure, resulting in the loss of carbon to the atmosphere

Progressive farmers realise the need to adapt their farms away from chemical-based conventional agriculture, where each pest, disease or fertility problem is 'zapped' with a chemical input. The first step is to acknowledge the complexity and interconnectedness of the farm system and the ecology of the land.

The following are some suggestions for actions to help regenerate your land.

Mob Grazing

Mob grazing is rotational grazing but with longer rest periods for the pasture in between. Once a significant mass of vegetation has grown, a small area of land is heavily stocked for a short period of time – for example, forty sheep might be given 0.2ha (0.5 acres) for two or three days before moving on to the next quarter hectare. This system requires good fencing or well-maintained electric fencing and a mobile water supply pipe and water trough.

Giving the pasture longer rest periods helps to control parasites such as intestinal worms. Mob grazing also allows the plants in the pasture to build up reserves and develop their root systems. A lot of the excess vegetation will be trampled back into the ground and seem to be 'wasted' but this is feeding the soil organisms and building soil organic matter over time. The intensive manuring of the pasture from the 'mob' stimulates fresh regrowth of the plants.

Allowing the pasture to grow for longer is beneficial to insects, birds and creatures in the soil, and fits into a system of mob grazing. The pasture regenerates in summer as it can flower and set seed. Birds, insects and small mammals can regenerate in turn because they are given space to breed and feed.

THE BENEFITS OF HERBS

In regenerative farming, herbs can be sown as part of grass seed mixes. Herbs can also be added by scratching up the pasture with a harrow when the grass is very short, broadcasting the seed and rolling the ground to press the seed into the soil. Herbs add minerals and antioxidants to the sheep's diet. Deep-rooting herbs like chicory break up the soil layers further down, improving aeration and drainage of the soil layers. Chicory is also a natural anthelmintic (antiparisitic), meaning that the tannins in the plant can suppress the development of intestinal gut worms inside the animal.

Agro-Forestry and Sheep

Agro-forestry can be as simple as regenerating hedgerows by coppicing, cutting back, double fencing and planting up gaps with new trees (*see* Chapter 1). This is satisfying work in the depths of winter and will keep you warm and very fit. When well managed, existing hedgerows and copses on the farm can yield useful fodder (leaves) for the animals and firewood or poles and sticks for use in the garden. More sophisticated agro-forestry systems incorporate the growing of trees with food and/or timber production, like the old traditional English orchard. The trees need protection from grazing animals to begin with, but once established, they can provide both fruit and fertile ground for growing fodder for sheep, poultry and cattle. The tree roots maintain good soil structure and attract beneficial insects, and the trees give shade and shelter to the livestock.

Systems of sheep production that incorporate trees with pasture help to mitigate against the effects of climate

Some plant species will supplement the sheep's diet with their leaves or provide the farmer with a crop of nuts, timber, or fruit. Hazel (pictured here) provides nuts, leaves for fodder, sticks for the garden and wood for making charcoal, and is a good species for coppicing.

change. Trees help to manage the flow and storage of water through the soil profile.

The main barriers to establishing an agro-forestry system are the need to protect the trees from the grazing animals while they are young and to meet their need for water. Some farmers are establishing lanes of trees in fields, meaning tractors and machinery can still harvest hay or crops in strips alongside the rows of trees.

ORGANIC SHEEP FARMING

The meat or milk from a sheep can only be sold as organic if the animal is born on a holding where the land is certified as being managed organically. The land must undergo a two-year conversion period to organic management where no herbicides, pesticides and artificial fertilisers are used, and the animals must be kept according to organic principles and fed an organic diet. The farm must hold a certificate from one of the licensed inspection bodies to confirm that it has been inspected and licensed to sell organic produce. It is illegal to sell food as organic without a licence. The inspection, auditing and certification of organic food gives the consumer confidence that they are buying food that's produced to the highest environmental and animal welfare standards.

In practice, it may not be worth a smallholding enterprise to pay the annual licence fee, but many smallholders still choose to farm to organic principles. If the smallholder's produce is sold locally direct to the consumer, they can build a relationship with the customer and explain how the food is produced.

Organic farming goes beyond just looking at physical and financial performance. Wider principles of organic sheep production include farming in a way that does not damage wildlife habitats on the farm or degrade the soil of nutrients, carbon and living creatures. Organic farming recognises that the health and well-being of soil, plants, animals and humans are interconnected. Soil fertility is protected and enhanced by crop rotations, rotational grazing, the use of nitrogen-fixing crops and green manures and by returning compost and animal manure back to the earth.

Red and white clover are commonly used by organic farmers to 'fix' atmospheric nitrogen into the soil.

Choice of Sheep Breed

Organic principles encourage the use of breeds or cross-breeds that have been developed to suit local climatic conditions as well as the intended market – for example, Suffolk sheep for rich, level, lowland pastures, and Welsh Mountain sheep for the tough conditions and thin soils on the hills and mountains. The use of breeds well adapted to local conditions is done for efficiency and animal welfare. For example, keeping a lowland breed such as a Dorset on the side of a Welsh mountain would require supplementary feeds to be bought in. Rough weather conditions would also put unfair stress on sheep that are bred for fertile fields with good hedges for shelter.

Organic Feed

As sheep are herbivores, organic standards require that sheep are fed a diet where more than 60 per cent of their dry matter intake is fodder from pasture or dried forage (hay or silage). The nutritional needs of the sheep should be met with homegrown fodder and only a limited portion bought in. An organic sheep farm should not be stocked too heavily, so that adequate fresh pasture is available to every animal.

Poll (hornless) Dorset ewes with lambs at foot. This is an example of one of the larger British lowland breeds, popular worldwide.

Sheep Welfare

The welfare of animals is assessed at the annual farm inspection, where problems and concerns can be discussed and solutions found to health problems. Organic sheep farms must have a written sheep health plan, which is adapted and updated annually. The principle of keeping sheep organically is that prevention of health problems is better than cure; the production system is therefore designed holistically, with the animals' welfare being as important as the financial and physical performance of the livestock enterprise.

The sheep are allowed to interact as social groups and are free to exhibit natural behaviours. Sheep should have access to pasture most of the year except when brought in for lambing or when weather conditions dictate their welfare is better protected by bringing them into a barn. Organic standards stipulate minimum space requirements for housed sheep (*see* page 39). In order to minimise stress and competition for feed, each individual sheep is given more space than is usual in conventionally farmed flocks.

Sheep Health

When health problems do arise, organic farmers are encouraged to question if a disease or illness can be prevented from recurring by tweaking the way the sheep are kept and looked after. Organic farmers differ in their approaches. Some use homoeopathy and added herbs to the diet to improve overall health, while others will incorporate vaccinations.

Anything that is used to treat or immunise the sheep must be detailed in the health plan and medicine book.

When there is illness in the sheep flock, affected animals must be treated and cared for as soon as possible and veterinary advice sought if required. There is a common misconception that organic farmers cannot

Vaccinating lambs against clostridial diseases using an automatic refilling vaccinator. Organic farms must include all routine treatments and vaccinations in a flock health plan that must be approved by the organic certification body issuing the organic licence.

use antibiotics and other conventional drugs. Where these are needed to improve an animal's welfare, veterinary drugs can be used under the guidance of a veterinary surgeon. The farm's medicine records must record the ID of the animal(s) treated, the medicine name and batch number, the date and the withdrawal period (in organic standards, the meat or milk withdrawal period must be double that stated on the product). Of course, organic sheep farmers should not rely on antibiotics and other drugs to routinely manage the health of the flock.

Sustainable farming is not just about land management. Allowing young people to develop a love and interest of the land is vital as well.

CHAPTER 4

ANIMAL WELFARE

Before buying sheep, ask yourself if you will have the time to look after them properly. Sheep will need gathering regularly in order to carry out routine health and welfare practices (such as footbathing).

THE FIVE FREEDOMS

In England in the 1960s the concept of the 'five freedoms' was developed by a committee of vets, agriculturalists and other specialists advising the UK government. It became formally known as the Farm Animal Welfare Council in the 1970s. The five freedoms were used as the basis for updating and writing legislation to cover the welfare of farmed animals. Laws on animal welfare were developed according to the following basic code.

Farm animals should have:

- Freedom from hunger and thirst – by ready access to fresh water and diet to maintain health and vigour
- Freedom from discomfort – by providing an appropriate environment, including shelter and a comfortable resting area
- Freedom from pain, injury or disease – by prevention or rapid diagnosis and treatment
- Freedom to express normal behaviour – by providing sufficient space, proper facilities, and company of the animal's own kind
- Freedom from fear and distress – by ensuring conditions and treatment that avoid mental suffering

Sheep should be checked daily.

When the ground is carpeted in snow, sheep are usually hardy enough to stay outdoors if they have plenty of fresh dry forage to eat, such as hay or silage.

Use the five freedoms as a basic guide for designing a system for looking after your animals. The code can help to ensure you meet all the needs of your animals and don't fall foul of animal welfare legislation. Happy animals that are content in their environment and have enough food will also be healthy, productive animals.

SHELTER

In adverse weather conditions, such as torrential rain, wind, snow and hard frosts, sheep should have adequate shelter and feed. If it's a barn, the indoor space should meet the minimum space requirements for each ewe for lying area and feeding (1.5sq m/16sq ft in conventional flocks, 2.5sq m/27sq ft in organic flocks). Shelter outdoors can be walls and hedges or the edges of woodlands – all of which can significantly cut the wind chill factor – but the sheep must have an area of ground that is sheltered from the wind and is not waterlogged where they can lie down comfortably.

Sheep can stay outdoors in frost and snow if they are not in danger of being buried or stuck in snow drifts and they have plenty of dry forage and feed to eat.

PHYSICAL HAZARDS AND ACCIDENTS

Just as basic good housekeeping can keep a home or workplace safe to live in, the same principle applies to a farm or smallholding. Broken gates, piles of rubbish, old or loose fence wire, collapsed fences, sharp objects, plastic waste – all pose hazards to sheep as they can easily become entangled, trapped or injured. Sheep are at risk of colliding with objects, particularly when being chased by a dog or human, and can hurt themselves on waste or old wire hidden in the grass or hedgerows.

Sheep will often put their heads through wire fences looking for leaves to browse on and get their heads stuck. Sheep must be visually checked at least once per day.

In the winter, when they have longer wool fleeces, sheep are adept at getting caught in brambles. They go into bramble patches to eat the leaves and the spikes attach to their wool like Velcro. The sheep will try to free itself by turning around but then becomes tied up, and if it is not discovered and freed, it will eventually die.

Getting stuck on their backs is another hazard. This usually happens in winter when the sheep has a heavy fleece and is pregnant: it rolls over and then cannot get up. The sheep cannot pass gas so it

A sheep that inflated with gas while stuck on her back. If left too long, they can die.

A young sheep caught in brambles. On land where brambles are growing, it's important to do a head count every day or check the boundaries of the land to make sure that a sheep trapped in a hidden part of the field is not missed.

In winter, ewes can get stuck on their backs and need help by rolling them over and supporting them until they regain their balance.

slowly inflates. If undiscovered, a sheep in this situation can die within twenty-four hours.

For all these reasons, sheep should be checked in the field every day.

ILL HEALTH

Sheep are known to succumb to illness very quickly once an infection sets in. Ensure your sheep are checked every day (preferably twice a day). The sooner an animal is identified as ill and treated, the more chance it has of survival. Look out for the following signs of ill health in a sheep:

- Separating itself from the rest of the flock
- Not grazing or eating
- Not moving when approached
- High temperature
- Severe lameness
- Wool falling out or dark, discoloured wool
- Stuck on its back or lying flat on its side
- Stumbling or weaving from side to side while walking
- Having trouble standing
- Coughing persistently
- Looking lethargic

FEVER SYMPTOMS

The normal temperature range for a sheep is 38.3–39.9°C (100.9–103.8°F). If the temperature is outside of this range, it is best to seek veterinary advice as the sheep may have a bacterial infection or another serious ailment that requires professional diagnosis. If a sheep is lethargic, not eating and drinking or finding it difficult to move, this can indicate a high fever. Move the sheep home to the yard, and take its temperature – put the tip of the thermometer inside the sheep's rectum. Consult a vet as soon as you can, because sheep can quickly deteriorate.

Sick Bay Facilities

Ensure you have a spare sheep pen with straw bedding ready in a barn or outbuilding to serve as a sick bay if you need to separate a sick sheep from the flock in an emergency; keep the animal in a pen for easy repeat treatments if necessary. With infectious diseases (such as contagious footrot), separating the affected animal from the flock will limit the spread of the illness to other sheep. Note that sheep do not like to be on their own, however, so provide a companion animal where possible.

Quarantine Bought-in Sheep

Quarantine new sheep brought onto the holding for twenty-one days if possible and footbath twice during the first week. You should observe them for foot trouble and signs of sheep scab, and treat for anthelmintic-resistant internal parasites. If keeping an accredited high-health status or pedigree flock, consult your vet about other quarantine screening. There are some long-term chronic viral diseases of sheep (known as 'iceberg diseases'), which can limit productivity and cause ill health in a flock, so your vet might advise screening tests.

PREDATORS

In Britain, mature sheep are fortunate that there aren't any large wild predators that will catch and kill a healthy adult sheep.

Crows

Crows can pose a threat to a ewe if she gets stuck on her back or has difficulty lambing outdoors. Crows will peck out the sheep's eyes while it is still alive. Crows can also attack newborn lambs.

Foxes

Foxes will occasionally kill and eat young lambs, usually within the first five days after they are born. Once a fox successfully takes a lamb, it will usually return for more. Sometimes it is an old or injured fox that cannot hunt effectively, or a hungry vixen that has cubs to feed. Weak, hungry lambs that are not getting enough milk are more at risk, as are twins with young, inexperienced mothers. A ewe can protect a single lamb from a fox but will struggle to defend two lambs, especially if there is more than one fox attacking.

If foxes are a problem, one solution could be bringing the flock closer to the house or into a barn at night. Foxes are also habitual and will usually approach the sheep field from the same direction each night, so if you know where the fox is coming from, three electric fence wires installed close to the ground on its path can be a deterrent (first wire 5–10cm/2–4in above ground, second and third wires at 15cm/6in intervals above). Foxes with territory in urban areas tend be bolder and less timid than rural foxes and can cause more problems. Foxes can be shot legally in the UK if it's done with the landowner's permission by a competent, trained person who holds a firearms licence. It is illegal to shoot foxes near a public footpath (closer than 50yd/m), where houses are close by, from a road or in an urban area. Extreme caution must be exercised if somebody is using a rifle, as the bullets can travel up to a mile.

Dogs

Dogs are descended from animals that were predators of sheep. Farming newspapers report attacks on sheep by domestic dogs almost weekly because dog owners fail to keep dogs on a lead when walking near sheep. This is a particular problem for sheep farms near towns and villages, as urban dwellers are often uneducated about the damage their dog can do.

Once a dog unearths an instinct to run after sheep, it is difficult to control and, given an opportunity, will wander off looking for sheep. A dog chasing sheep can cause enormous distress and damage to a flock, particularly if the sheep are pregnant or in the breeding season.

A stray dog unattended will usually chase a sheep until it collapses and then start biting it in the throat until the sheep dies. Even if a dog doesn't get this far, the stress of being chased by a dog can damage the sheep's health.

If possible, keep dog-walking areas away from sheep grazing, not just because of the risk of attack, but to prevent disease. Dog faeces commonly carry *C. tenuicolis*, the larval form of a tapeworm species. Infected dogs pass tapeworm eggs in their faeces and the larvae damage the liver of infected sheep and can cause lambs to abort.

SHEEP STRAYING

Sheep require good perimeter fencing, as it won't take them long to find places where the fencing is inadequate and allows them to escape. People leaving gates open by mistake is a favourite route to freedom. Public footpaths are best served with a stile to climb over a boundary, or a kissing gate.

Domestic dogs are a real threat to sheep if not kept under control.

Electric fencing can form an effective temporary boundary but must be checked daily.

Primitive and hill breeds of sheep are small and nimble with a considerable wanderlust, and require secure fencing at least 120cm (4ft) high. Some are even adept at climbing over dry stone walls. Stray sheep are at risk of harm from encounters with people or dogs chasing them, and on busy roads there's also a possibility they could cause a road traffic accident.

SHEEP IDENTIFICATION

As well as applying the statutory ear tags, it can be a good idea to apply a paint mark in the same colour and shape to every sheep in your flock. If the sheep do then temporarily disappear, they can be easily identified even if they have lost their ear tags.

Hill farms using common grazing rights traditionally have 'ear-notching' pliers to cut a small part of the edge of each sheep's ear. Each farm in the district has its own shape so the farm of origin of any wandering sheep would be known by every shepherd in the area. During the annual gathering times on the hills, such as shearing time, the sheep can be separated into flocks according to their ear notch pattern.

Ear Tags

Ear tags are a statutory legal requirement in the United Kingdom. Each tag contains a UK flock number, which is assigned to the farm (holding) the animal was born on. The ear tag will also carry an individual number. Breeding sheep are required to be double-tagged – that is, have one tag in each ear, one of which has to be an electronic tag known as an EID tag.

Check the UK government website or those of the devolved administrations in Wales, Scotland or Northern Ireland for the latest rules on registering as a sheep keeper, tagging sheep and how to record sheep movements.

Lost Ear Tags

Unfortunately, sheep sometimes catch an ear tag on a fence or hedge and rip it out, causing injury to the ear. The ear usually heals well and doesn't cause the sheep any trouble once healed, but it will be left with a damaged ear, spoiling its looks for life. If you discover a

Distinctive paint marks, such as the blue number or green mark on the head shown here, can be useful for identifying your sheep if they are found straying away from home.

The yellow EID tag contains a microchip that can be read by the farmer, auctioneer, abattoir worker or welfare officer using a stick reader or wand. The numbers can be transferred to a phone app or computer via Bluetooth.

recently injured ear, spray it with antibiotic spray (from the vet) or an antiseptic spray. Check for signs of infection such as yellow exudate (pus) or weeping liquid, and in midsummer apply a fly repellent until it has healed.

Animals that have lost their ear tag(s) should have a replacement tag applied. If the individual identity number of the sheep is known, a replacement can be ordered with the sheep's own number on it. If a sheep has lost both ear tags and its unique number is not known, a red replacement, with the farm flock number on, should be applied.

RECORD KEEPING

In the UK, it is a statutory requirement to keep a holding register on paper or electronically that records movements onto and off the holding. The register must also record births, deaths, ear tags purchased, medicines purchased, medicines used and medicine withdrawal periods.

Recording Sheep Movements

Each nation within the United Kingdom will have its own rules for reporting animal movements, so check the relevant government website for the movement reporting requirements. Records can be kept on a computer, on paper, or on a farm software app on a mobile device.

Veterinary Medicines

If a medicine is used to treat an animal, the treatment date should be recorded along with the animal identity, the length of treatment, reason for treatment, name of the medicine, medicine batch number and expiry date, and the withdrawal period.

Medicine Withdrawal Periods

The withdrawal period is the length of time after treatment with a veterinary drug, either

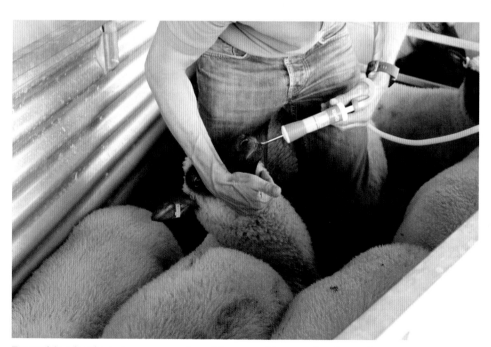

Drenching lambs using a secure handling pen. To meet legal requirements, you must keep a record of any medicines given to the sheep.

SUMMARY OF ANIMAL WELFARE

- Familiarise yourself with the five freedoms and analyse whether your grazing area, fencing, buildings, labour and time availability will allow you to keep a contented and happy flock of sheep.
- Create a safe and secure environment for your sheep where they have enough land to feed and exercise.
- Do not allow the sheep access to physical hazards such as waste piles and working yards. Make sure the grazing area is free from rubbish and loose string, plastic, wire and machinery.
- Provide shelter in case of adverse weather conditions and protect the sheep from predators and domestic dog attacks.
- Check your sheep every day and treat any ailments promptly, using an indoor pen as a sick bay if necessary.
- Quarantine bought-in sheep before introducing to the rest of the flock.

records. Withdrawal periods are calculated by regulators according to the rate at which the drug is metabolised inside the animal and the length of time it takes before any residues of the substance are deemed to be at safe levels. In certified organic farming, the farmer must double the length of the standard withdrawal period for any treated animals.

FARM ASSURANCE SCHEMES

Farm assurance schemes inspect farms to ensure that the farm meets basic requirements in looking after livestock, including adequate facilities for handling and moving livestock. A farm assurance audit will also look at medicine usage and movement record keeping, and requires the farmer to have a basic flock health plan developed with the help of a vet.

Good returns can be made by selling batches of quality fat lambs at an auction market or direct to an abattoir that supplies a supermarket. To access these markets for selling batches of lambs, you may need to have your holding inspected and pay for a farm assurance certificate; check with your local market or abattoir what their requirements are. You may also be able to sell your sheep to a larger farm.

orally or via injection, when an animal should not be sold as meat for human consumption. The milk from the treated animal should also not be used. It is important to note the length of the withdrawal period in your

CHAPTER 5

SHEEP BREEDS OF BRITAIN

A lineup of rams being judged at the annual Llanwenog Sheep Society show and sale. Llanybydder Mart (2022).

Llanwenog sheep, Ceredigion, west Wales. Each region of Wales, England and Scotland has its own distinctive sheep breeds, as well as many continental breeds and cross-breeds.

The importance of the humble sheep to the agricultural and economic history of Britain is demonstrated by the incredible array of native regional sheep breeds found here. There are over seventy breeds of sheep native to Britain plus myriad cross-breeds. They range from primitive, hardy, 'unimproved' sheep, such as the Soay and Shetland, to hardy mountain and hill sheep like the Swaledale and Welsh mountain breeds.

THE RIGHT BREED FOR YOU

Look at the breeds and types of sheep closely. All sheep breeds come with their positive attributes and limitations. Look for a type to suit your land, facilities and management system. Your lifestyle and the time available to dedicate to sheep is also an important consideration. For example, a pedigree breeding flock of ewes will require regular year-round work, while keeping retired breeding ewes as lawn trimmers or for wool production would be less effort. Keeping weaned 'store' lambs bought in and only kept for part of the year before being sold for meat or put in the home freezer would also be less commitment than having a breeding flock. Do you have time for sheep at all? If the sheep keeper is not fully committed, the animal's welfare will deteriorate.

Sheep markets and agricultural shows are a wonderful way to get to see plenty of different breeds and to talk to the breeders before deciding which type of sheep to buy.

THE RIGHT BREED FOR YOUR LAND

If you are high above sea level on a windswept hillside in the west or north of Britain, the sheep you keep need to be tough and hardy. Along with aspect and altitude, consider soil and forage quality too. How much grass can be grown naturally to feed your flock? Mountain soils are often thinner and more acidic. When these soil conditions are coupled with a shorter growing season (in the north or at altitude) and colder, windier conditions (in the west), annual forage production is a lot less than in the sheltered lowlands. In these conditions, only smaller, hardy hill breeds will perform well (without buying in feed and housing sheep for longer periods).

Organic standards encourage the selection of native breeds adapted to the local climate. How hardy do your sheep need to be?

If you are unsure about the quality of your land (in food production terms), you can consult a UK soil map, where soils are classified into different grades according to soil depth and fertility and their crop-growing potential. Look up your region to get an idea of your land type and quality.

For the purposes of subsidy payments, UK land was also classified by an EU system to designate areas where farming conditions are considered difficult:

1. Severely Disadvantaged Area (SDA)
2. Disadvantaged Area (DA)
3. Less Favoured Area (LFA)
4. Non-LFA land

If you are situated in the first two areas, it is likely that a hill-type sheep or primitive breed may be the most suitable.

Welsh Mountain sheep. Most sheep breeds have a dedicated breed society that organises a show and sale where sheep farmers can buy and sell ewes and rams.

Examples of native breeds and the terrain they are suited to

Terrain	Breed type	Examples
Hills and uplands, coastal plain, moorland, marsh	Hardy maternal breeds	Scotland: North Country Cheviot, Scottish Blackface England: Herdwick, Masham, Swaledale Wales: Welsh Mountain (various types), Black Welsh Mountain, Brecknock Hill Cheviot Primitive lightweight: Soay, Manx Loghtan, Hebridean
Fertile lowlands	Heavy lowland sheep	Suffolk, Dorset, Hampshire Down
Good grass-growing areas (where arable farming is still challenging)	Efficient commercial grazers	Lleyn, Llanwenog, Clun Forest, Romney, Welsh, Scottish and English types of mule cross-breeds

Herdwick sheep, one of the many hardy hill breeds of the north of England.

Young lambs (under four weeks) bred from Welsh mule and Lleyn ewes from a Charollais ram. The Charollais is a French meat breed commonly used in Britain as a terminal sire.

Poll Dorset ewe lambs for sale in the autumn.

MAKING MONEY FROM YOUR SHEEP

Sheep keeping is good for land management, provides a satisfying pastime, and preserves a link to traditional rural living, but once you have chosen a breed or type that suits your land and lifestyle, how can the sheep help to pay the vet and feed bills? The choice of sheep breed will influence which markets you can access.

Decide on the scale of business you would like to pursue and the time you wish to commit. Is it going to be a full-time business to make a living from or the extension of a satisfying hobby with the aim to recoup some funds? Selling artisan products can be difficult because it is impossible to compete on the price and production efficiencies of mass-produced products. Can you produce and sell a desirable product at a price level that means the customer will return?

The most obvious saleable products from sheep are meat, wool and breeding stock.

Meat Sales

Sales can be as simple as taking live lambs and older ewes and rams to market or selling fat lambs direct to an abattoir. Returns will be better if using types or breeds of sheep that have good 'conformation' (size and shape). Abattoirs generally want lambs that are 40–45kg (88–99lb) liveweight, which translates into 17–22kg (37–48lb) dead carcass weight. A medium to large well-fleshed type of sheep is more suited to this market, as payments for a lightweight carcass

of pure-bred small and lean primitive or hill breed would be low. Bear in mind that larger breeds will not always be more profitable once costs are deducted, however.

Introducing hardy sheep genes into the breeding flock can keep health status high, reduce vet and feed bills and improve the efficiency with which the animal converts grass into meat. This is the reason why many lowland farmers have traditionally favoured the cross-bred 'mule ewes' that are reared on the hills.

Direct Selling

Selling direct to a restaurant, wholesaler or your own retail customers can be an option. Farmers' markets, independent butchers and online sales all offer opportunities. It is hard work and a tough market with small margins but direct selling can offer a sustainable long-term market for your products – and good job satisfaction if you enjoy being busy, creative and sociable.

A commercial cross-bred ewe produced from a Welsh Mountain ewe and a Blueface Leicester ram. Thousands are bred every year on the hills and sold to farms in the lowlands.

The marketing and production plan must be based around a desirable product of the right age and fat cover, with excellent presentation and freshness, and should include a story of the provenance and effective hassle-free distribution to the customer, as well as the all-important flavour and tenderness of the meat. Investment may be required in your own facilities for butchery, packing, labelling and distribution. Accessing a local small abattoir can be difficult and sourcing excellent butchery skills can be problematic unless you are willing to learn yourself. There are many farm shops, online outlets and independent butchers making a success of direct selling into local markets, so see what they offer before taking the plunge.

Wool Sales

Breed choice will influence the quality and fineness of the wool produced. Sales of wool can be as simple as selling raw, unwashed fleeces to local craftspeople (*see* Chapter 18). Demand and prices will be low, however, unless the wool is pre-washed before sale or of superior quality. Turning your homegrown wool into finished products and selling goods online can be a satisfying small enterprise but does require plenty of time and good craft skills. The British Wool Marketing Board classifies fleece into many different grades linked to the breed of origin. Have a look at the grading system for information on which breeds might produce the kind of wool you are looking for. Research independent small-scale wool processors and gather costings and production information, comparing it with what you would be able to do at home.

Sheepskins

Small abattoirs will usually allow the keeper to have the skins back, but they will need immediate processing and curing to prevent them from going off. If done at home, the sheepskins would be a wonderful product to sell locally or to friends and family. There are a handful of independent tanneries that can process sheepskins but production costs will be high; it is difficult to compete with the mass-produced wool rugs on the market that are produced in New Zealand and Australia, processed in China and sold in the UK.

Breeding Sheep Sales

Ewe lambs that are not needed to add to the home flock can be sold for future breeding stock. For higher prices, keep them for a year and then sell them as well-grown yearlings. This does require sufficient grazing over winter, though. Surplus male lambs can be sold for meat for extra income; they will fatten well and be easy to keep if castrated shortly after birth.

Ram lambs kept 'entire' can be reared and sold for breeding, but only the highest-quality rams will be in demand. Selection needs to be careful, as there is usually a surplus of rams on the market.

Important traits to look for are conformation, growth rate, weaning weight, appearance, wool quality, jaw and foot correctness and the way a ram walks and

A prize-winning Lonk ram in the sale ring. Lonk sheep originate from the Pennines range in northern England.

Only healthy ram lambs with a good growth rate and correct breed points should be selected to breed from. This is the Llanwenog champion ram lamb 2022 from Elfyn Morgan's flock.

stands. Also, the performance, health, vitality and appearance of the dam should influence the decision on whether to keep a male lamb as a ram or not.

Keeping a breeding flock of ewes and selling the surplus lambs is a clear way to get some returns on your flock, but good prices will only be achieved if your nucleus breeding stock is of good quality. A known pedigree breed that is in demand will achieve better prices. Whatever the breed, sheep must be well looked after, with all vaccinations and health treatments up to date.

If you are dedicated and willing to travel, the showring can be a good place to showcase your pedigree flock.

Grazing Services

In areas of the country where sheep numbers are relatively low, it may be possible to get patches of free grazing for short periods or rent sheep out for conservation grazing or to manage the pasture. A good trailer and mobile set-up of hurdles would be required, and likely electric fencing kit. The main challenge would be fences that have not been maintained and supplying water to the sheep, plus the distances you would need to travel to check the sheep.

CONSERVATION OF RARE BREEDS

If you could have looked at Britain from the air before the industrial food era you would have seen an intricate patchwork of fields, gardens, woods, heathland, common grazing, moorland and mountain. Dotted within this rich and diverse landscape were sheep. Each region kept a different type of ewe and each part of the terrestrial patchwork was grazed by a sheep breed that suited the local weather and the needs of local markets.

Fortunately, in the 1970s, a group of forward-thinking ruralists realised that to lose such a rich genetic diversity of farmed livestock breeds in the UK would be pure folly, and in response they founded a charity called the Rare Breeds Survival Trust (RBST). The mission of the RBST is to conserve the rich diversity of British breeds

A Manx Loghtan ewe, suitable for conservation grazing.

of farm animal and to show the economic, social and environmental relevance of native breeds.

Today, across Britain, the work of conserving and improving the myriad types of rare breed livestock continues. There is an unsung network of traditional grassland farms. Shepherds with small flocks and herds, hobby farmers and smallholders, organic farmers and local food marketing initiatives, box schemes and farm shops: all work to produce high-quality food of known provenance and to keep alive the old breeds and sustainable farming methods.

THE COST OF UNIFORMITY

In the latter part of the twentieth century, the growing popularity of supermarkets led to the centralisation of supply and the mass transportation of foodstuffs over long distances. To feed this new industrial food system, and fuelled by EU subsidies, farms and fields became bigger and more specialised in what they produced. This was the case for both crops and livestock. Farms with fields that could be cultivated for crops grew fewer varieties of plant, while livestock farms concentrated on producing only the type of animal that would fit the supermarket's desired specification.

The drive was to persuade farmers to produce animals of a uniform size, fat cover, shape and age. The different cuts of meat were no longer sold fresh from the butcher's counter and cut with the craft of a skilled butcher in the presence of the customer. Carcasses were now cut up in a factory and sold in standardised small plastic packets. The appearance and convenience of the product were and are more desired than taste and provenance. Breeding changed dramatically to meet the demands of the factory production line. Uniformity was the name of the game and many of Britain's regional breeds did not fit the narrow specification of what a sheep should be, so their numbers quickly dwindled.

A Portland ram, one of the many breeds that the RBST helped to rescue.

SHEEP BREED SUMMARY

- Sheep breeding is a bit like cooking – there are so many ingredients to choose from and endless regional styles and ways of mixing the ingredients and bringing them along to produce a finished dish.
- Choose a breed, cross-breed or type of sheep that will suit your land and lifestyle. It should be a sheep that you like the look of and that has a temperament you can get along with, as you will spend a lot of time together.
- There are many ways you can use your sheep to raise money, such as pedigree sales, land management services, fat lamb sales, buying lambs to grow on and sell, wool production and direct selling of meat.
- Larger lowland or 'down' breeds have been bred to produce the best and highest value meat carcass but will need more feed and require more care than hill breeds. Many commercial farmers use cross-bred 'mules' for lamb production, as they combine the hardiness and mothering ability of the mountain sheep with the size and meat quality of lowland sheep. Direct selling products from your flock and adding value by home processing is possible but requires thorough research of the market, the production process and pricing.
- Commercial considerations aside, the conservation of a rare breed, the maintenance of traditional wildlife rich pastures and field boundaries and the picturesque nature of sheep grazing within a landscape can be reason enough to keep a few sheep.

CHAPTER 6

MOVING AND HANDLING SHEEP

Gathering and catching sheep when you are inexperienced can be a real challenge. If possible, gain some practice working with an experienced shepherd.

A large flock of hardy Welsh mountain ewes. Gathering a large flock would usually require a well trained sheepdog and an experienced shepherd.

CATCHING SHEEP

Learning how to catch and handle sheep efficiently without causing distress to the animals or yourself is one of the most important skills to learn when starting a new flock.

Without adequate preparation and facilities, catching sheep can be one of the most frustrating endeavours known to humankind. Being evolution's prey animals, sheep are adept at avoiding what they see as a predator's advances.

For sheep that are not bucket trained or hand reared on a bottle from birth, the natural inclination is to view their master as trouble, and to try to run away from them. Depending on the sheep's temperament, age and the time of year, the reaction to seeing a human can either be a wild dash to the other end of the field or a reluctant shuffle away, just enough to stay out of reach. If it is winter and they are hungry, the sheep will instead run towards a person if they think there is feed.

It is rarely a good idea to try to catch hold of sheep in the field unless they are very tame. Once a sheep knows it is being targeted, the adrenaline makes it hyper alert and able to avoid your advances.

Without experience, even people who are old enough to know better, will, in the heat of the chase, fling themselves headlong into a rugby tackle to catch a sheep. If you are not a proficient rugby player, this will not often go well for you, and if you are a good rugby player, the outcome will not be good for the sheep. Fortunately, there are better ways of catching sheep.

Once your sheep have become accustomed to the farm and the layout of the fields, hills, gateways, natural obstacles, puddles and fences, they will be much easier to move around your patch.

A solution to difficulties gathering sheep could be as easy as taking them in on a different route to avoid a big puddle in a gateway or asking your well-meaning assistant not to stand directly behind the pen that you are trying to get the sheep into.

If the sheep are hard to catch in the pen, try moving them into a smaller pen so they are easier to catch hold of without chasing them around and around.

Catching an Individual Sheep

Sheep seem to know the stockperson's intentions even before they have been

SHEEP SENSE

Sheep sense is a mysterious sixth sense which you will need to master if you are to handle your sheep efficiently and with the minimum of stress for you and the flock. Through experience you will get to know how the sheep will behave in different situations. Getting to know your sheep's behaviour and their quirks and anxieties will help you work with them more effectively.

revealed. If your aim is to catch an individual from the flock because it is lame, the lame sheep seems to know instantly that you are looking at it and planning to single it out from the group. Even before you move a muscle to try to catch it, it will be on the alert and move away, becoming impossible to catch. In a small flock, it is sometimes possible to get the sheep eating out of a bucket and then grab hold of it while it is distracted. The problem with this approach is that once the sheep is caught, you won't have the tools and equipment or medicines at hand to treat an ailment, and the next time you need to catch it, the sheep will be more wary.

In larger flocks without bucket-trained tame sheep, it is usually impossible to catch an individual animal in the field. The whole flock must be gathered into a secure handling area where the individual can be singled out and treated. It rarely pays to cut corners and try to catch sheep without proper secure pens set up in advance.

GATHERING THE FLOCK

When gathering the flock, you will soon know when is the right time to advance and apply pressure to get them to move forwards, and when you need to stand back a little, giving the sheep time and space to think. Apply too much pressure in a hurry and the sheep will run to the left or right with the sole purpose of escaping, and may not see the opening in the hedge or the gateway through which you are trying to guide them.

Sheep sometimes need time to think before a bolder one will make the first move and lead the others through the gate. If they won't adopt your idea of moving in a certain direction, then the shepherd can be in for a struggle!

Teamwork

If you are lucky enough to be working with somebody else, try to choreograph your moves so you are working together.

To the shepherd gathering the sheep, the gather is the start of an intense hour or so of important work that needs getting on with swiftly. To the assistant, being paid in cake or such like, it may be seen as a fun, novelty leisuretime activity.

Begin the job together if possible, so that your assistants don't arrive just at the moment you are panting and sweating having just gathered the sheep from the field on your own. It is very frustrating when you have the flock poised to run through a gateway and somebody suddenly turns up to 'help' at the critical juncture. As you wildly gesticulate and shout from the other side of the flock to indicate for the person to move out of the way, the sheep will usually have hot-footed it back down the field.

Lone Sheep

Within any group of sheep there will usually be a renegade or decoy who will try to disrupt proceedings by jumping through a hedge or getting stuck behind a gate as the others trot on merrily up the track. Often the individual can be diverted back to the group easily by the shepherd standing aside or by walking around the obstacle. If this is not possible, the rogue sheep can derail the whole operation. As focus shifts onto catching the stray, the others may try to follow the escapee.

It is tempting to get the group into the pen before going back for the runaway. In most instances, this is a bad ploy and where the shepherd tends to lose their head. Unless it is tame and has been bucket-reared, a lone sheep away from the flock will panic. By chasing it, the adrenaline rises as you and

Sheep can often be awkward to move through gateways. They do not like it if there is something or someone unfamiliar in their line of vision, or puddles or mud in the gateway. Sometimes they need time to think before one sheep will lead the others through.

If your friends or family are on board with the whole sheep keeping enterprise then all the better for you. A flock of wild young sheep can sometimes take more than one person and a dog to catch.

the sheep go into the age-old predator and prey stand-off. The lone sheep pursued will go into flight mode, and there is a much higher chance that you or the sheep will get injured. In a blind panic, it may fling itself against wire fences.

A simpler solution to chasing an individual is to back off and let the runaway rejoin the flock by itself. Even if that means releasing the whole flock again into the paddock and regrouping, this is often the best way to ensure every individual ends up safely in a pen or barn.

Footwork

The knowledge of where to stand in the field or yard in relation to the sheep will come with experience. Unless the sheep are trained to follow a bucket of food, they will need to be herded or shepherded along from behind. If doing this with other people, co-ordinate your movements as the sheep will not want to walk towards somebody they do not know. A common mistake is to send someone to open a gate or block a track by standing in it; the assistant then stands in full view of the sheep, which get spooked and will not go in the direction intended.

If the sheep are not trained to run after a bucket of feed, and a sheepdog is not available, moving sheep can involve a lot of running about.

Over time, you will work out your own strategy for easily catching or moving the sheep when you need to. The aim should be to gather the sheep with as little walking or running as possible, with as few people as possible, so that you don't have to rely on others every time you need to round up your flock.

Walking along with a stick like a traditional shepherd and listening to the birds is very fulfilling if you are not in a hurry. It's a joy to spot the wildlife in the field as you walk round the sheep on your shepherds' rounds.

MOVING SHEEP

If the flock is greater in number than around twenty, or the sheep are not trained to follow a bucket of feed yet, it is almost certain that gathering sheep on foot will be necessary. The real fun and learning of how to move sheep is in the trying, but below are some common approaches.

PROPS TO GET SHEEP MOVING

- Rattles and rustles
- Something flappy, like a plastic bag or old coat, especially if the sheep are docile and not particularly scared of you.

Arm Extensions

Friends on a neighbouring farm to myself gave up on keeping a sheepdog a few years ago. They use long bamboo canes from the garden that still have the leaves on the end to 'guide' the sheep in the right direction as they walk behind them. This seems quite an effective tactic. You could try tying plastic bags to the ends of long canes as an alternative.

Voicework

If you have the pleasure of helping an experienced sheep person with moving sheep you may be surprised to hear a ditty of strange hisses, clicks, grunts, shouts and whistles from them. Sheep don't seem to mind you talking to them in a polite, reasonable way but dislike the more basic noises and will usually run away when hissed, grunted, whistled or yelped at. A loud whistle from a distance will usually get the sheep to flock together in the field. This puts them in flight mode so is only useful if the flock is being rounded up by a sheepdog or person on a quad bike.

Sheepdogs

A sheepdog is essentially a dog that wants to chase sheep, but one that you can control. A sheepdog should be able to go clockwise (come bye) and anticlockwise (away) around the sheep and stop and come back when you command it to. If it does not, it is best kept away from sheep, unless it is a young dog in training.

A good sheepdog is worth at least five people in terms of the ground it can cover and the 'flocking' effect it has on the sheep.

The advantage of a dog is that the sheep will flock tightly together for safety. The sheep also know that a sheepdog is faster than they are and those on the flanks of the group are less likely to try to escape.

Several times in the past I have gathered four extra people together to help move a group of sheep out of a field into the yard. The sheep gather near the gateway but then turn around to face their pursuers. A couple of brave sheep make a run for it straight past one of the humans and the rest follow. With a good sheepdog this will not happen, and I and one of my sheepdogs can move even the most reluctant flock without too much fuss.

Without a dog, the sheep spread out more, creating the possibility of a breakaway to the left or right as you walk up behind them.

Pet Dogs

There are many dogs that will chase sheep but there are not many dogs that are worthy of being called 'sheepdogs'. An untrained dog can do a lot of damage to the sheep. When chased by a loose dog, sheep will panic and run and can abort their lambs if pregnant.

It is possible to have a dog on a lead and to walk up behind the sheep to move them along. This will not stop the sheep from breaking left or right, however, and if the dog wants to go after them, you will be unceremoniously dragged along behind. Another disadvantage is that you're teaching a pet dog how fun it is to chase sheep without any control. This is very ill-advised, especially if said dog has a bit of border collie in its breeding. When you are not looking, it may well slink away from home to find the sheep and have an exciting time running them ragged.

Panic-stricken sheep chased by a dog will run into fences and injure themselves trying to escape. If chased by a dog for long enough, a sheep will eventually collapse from exhaustion and the dog may kill or injure it. Dogs and sheep were domesticated by humans thousands of years ago, but there is still a very uneasy relationship that is not totally removed from wolf versus prey.

Some types of dog will not show the slightest bit of interest in sheep. A dog like this can make a better pet on a sheep farm or smallholding than a border collie, collie cross or a larger dog with working genes.

Quad Bikes and ATVs

Large and small sheep farms often rely on a quad bike (or all-terrain vehicle) to save on the leg work, speed up checking the sheep and carry hay or feed into the fields. Quad bikes are often used instead of sheepdogs for rounding up sheep with the noise of the horn and engine frightening the sheep into flocking together and running away.

A quad bike can also pull a small trailer, which is useful if you have to bring a sick ewe and her lambs into the yard from a distant part of the holding, or for moving ewes with young lambs about. A quad can also be used to pull small machinery, such as grass harrows and flail mowers used for topping grass in the summer.

Overall, they are useful tools but not essential for smaller flocks or farms, and you will be fitter without one. I have learnt from experience that a stricken ewe fits quite snugly on her back in a wheelbarrow if no other transport is available to get her back to the barn.

Bear in mind also that quad bikes can be dangerous if the rider is not properly trained and they are noisy and polluting.

BUCKET TRAINING SHEEP

If the flock is fewer than around thirty sheep, bucket training can be a good option, though it does take some time to begin with.

The feed can be whole cereal like rolled oats or barley, sugar beet pellets or a concentrate feed in the form of a pellet or 'sheep nut'. Ensure there is enough space at the trough to allow all the sheep to eat at the same time.

To bucket train sheep, have a sheep feed trough in the field and take some concentrate feed out to the sheep in a bucket every day. Rattle the feed in the bucket a little as you walk before pouring the feed into the trough.

Take care not to exceed the recommended daily amount per sheep of cereal or concentrate, as this can cause a potentially fatal acidosis of the rumen. The sheep will also become very fat if overfed throughout the year and this negatively affects their fertility and easy lambing ability.

At the beginning of bucket training they might be very shy and show no interest but eventually they will become inquisitive and start nibbling at the tasty treats you have left them. If you can put the feed out at the same time each day to begin with, they will get into the routine of seeing you and will learn to expect a treat.

Sheep are quite particular and will not eat wet food. Early on, while the sheep are still getting used to being fed, it is best to leave only a small taster or sprinkling of food in the trough. If you have a covered feed trough it will help to keep the food dry for longer. Lambs or sheep not previously bucket fed will take a few days to become accustomed to eating from a trough. Clear up any uneaten wet food from the previous time before leaving more dry food out for them. Start with small quantities for the sheep to nibble at.

Feed from a bag or bucket is high-energy convenience food. Understandably the sheep will be keen for it. Ideally, though, they should be treated like the hard-working herbivores that they are and as much as possible be left to earn their subsistence by grazing and wandering. Too much bucket feeding will leave them standing by the gate calling for food rather than grazing. Once the sheep are running to you and the bucket they need not be fed every single day, unless daily feeding is part of their winter pre-lambing diet or the sheep are experiencing harsh winter conditions, like frost and snow or prolonged heavy rain.

In the summer and autumn the sheep should have plenty of grazing and don't need extra feed. A small amount of hard feed from a bucket once a week or once a fortnight will help to keep them tame and easy to catch. Once tame, the sheep will gallop towards you at the sight of the bucket – be careful not to get mobbed and knocked over. They can get very boisterous as they push and shove each other out of the way, and 60kg (132lb) of herbivore barging into the knees from behind has comical but potentially injurious consequences. If possible, get yourself to the trough and spread the food out before the sheep come hurtling over.

Place the troughs in drier parts of the field, so your feet do not slip or get stuck in the mud. If the sheep become too unruly at feeding time and they don't need the feed for pre-lambing nutrition then you could consider stopping feeding them completely. The sheep will remember what treats a bucket holds and still come running when you stand at the gate and shake it.

SHEEP PENS

Secure sheep pens and sheep races are required for carrying out routine tasks.

A good shepherd puts a lot of thought into the design and siting of sheep pens because a good set-up of pens can save a lot of stress and effort for the sheep and the shepherd.

When siting your handling pens or planning a route for moving sheep on foot, keep in mind the following:

- Choose a site that does not get waterlogged, preferably with a stone or concrete base. This will help with handling in wet weather and when dealing with feet.
- Sheep always tend to run uphill if they feel threatened, so it's easier to catch them in a corner at the top of a slope than at the bottom.
- Sheep do not like mud or standing water, so they can stall if there are puddles across a track or gateway. Putting clean stone in gateways can help.
- Sheep do not like unfamiliar objects, so they may be difficult to walk past parked cars, for example. A shiny car or person in the sheep's line of sight can 'spook' them. Sunlight reflecting off a car or window could also deter them from moving in that direction.

Sheep will be reluctant to enter a barn or stable if they have not been in there before, particularly if it's quite dark and gloomy inside. Lights can help to brighten the space. If the sheep can look all the way through to daylight on the other side of the barn they will be more willing to go inside.

Use a small pen for handling sheep when:

- Turning sheep over to treat feet or for dagging (clipping dirty wool off the back end)
- Injecting a sheep with antibiotic
- Tagging young lambs

A sheep pen and race made from timber post and rail fencing and galvanised metal gates with compacted stone floor. Decide if you need a fixed or mobile set-up of pens for sheep handling.

- Shearing – for holding the sheep before they are hauled onto the shearing board
- Loading sheep into a trailer

The following tasks are best carried out in a sheep race:

- Vaccinations usually given under the skin of the neck
- Using a drench gun to administer medicine
- Tagging mature sheep
- Dagging ewes while they are standing up
- Applying pour-on chemicals for lice or to prevent flystrike
- Drafting sheep, separating sheep into different groups
- Standing sheep in a footbath

Leading Sheep into a Pen

Once the sheep are following you and the bucket towards the pen, how do you shut them in? If operating solo, you will need a cunning plan of how to shut the gate, or hurdles, behind the sheep to trap them in. This is not always easy if the sheep are thronging around the bucket you are holding. If a feed trough is ready in a larger pen, the feed can be quickly scattered in the trough and the pen can be closed behind them to prevent break-backs and escapees while the sheep are eating. Planning the gather in advance and ensuring all the gates and hurdles are in the right position beforehand will make the job smoother and take the drama out of it. Sheep can be melodramatic little souls at the best of times. This is especially true for ewes with lambs at foot, who can be panicking and turning around in all directions looking for their offspring.

Before handling sheep in a pen or race, warm up your muscles with a few stretches, lunges and squats. Sheep handling involves a lot of reaching forward, bending from the knees and lower back.

THE SHEEP-HANDLING AREA

The less space the sheep have to move around once they are penned up, the easier it will be to handle them. It is also easier if you can handle a small group at a time. A pen that is just big enough for half a dozen ewes makes a good working space.

It is difficult to get the sheep into a small pen to begin with, as they will break back past you. A solution to this problem is to use some fencing as a funnel in the corner of a field or yard leading to a set of sheep hurdles in the corner, which can be closed behind the sheep.

Another option could be to run them out of the field, shutting the gate behind you as the sheep run up a lane or track and into a pen or building.

In any location, a large pen made from sheep hurdles, leading to a smaller pen of hurdles on the inside, makes the operation smoother and avoids individual sheep making a sudden break past you before you can close the hurdles. It can be useful to even have a third pen into which you can release the individuals that have had treatment.

If possible, have a metal hurdle or gate that is easy to open and close, allowing you to easily sort sheep from the group. Sheep hurdles that join with metal rods are generally sturdier and more secure than hurdles with metal loops.

It is possible to use hurdles to make mobile pens, carrying them to wherever they are required. This is time-consuming and heavy work, and the pen is never quite as solid as a semi-permanent fixed arrangement that has been built with care. Mobile pens are sometimes the only option, however, if keeping sheep on rented fields or far away from the main farm yard.

Having a well-constructed permanent handling area can save time throughout the year, as you will not have to build a pen every time the sheep need attention. Sheep can also sense when a pen is weak and flimsy so will do more to try and escape.

If using a temporary mobile pen, the external hurdles need to be fixed to something immoveable, like a fence post knocked into the ground, a concrete wall or gate post. Alternatively, use another hurdle set at right angles to the pen sides to buttress the arrangement. A group of ten sheep weighing 50kg (110lb) each, all trying to move in one direction can exert considerable force with their forty hooves and easily knock over a flimsy barrier. It is common for a novice shepherd to underestimate how robust a handling pen needs to be. If hurdles are likely to come apart when pushed, make sure they are tied securely at the top and bottom to a strong fence, gate or another hurdle. It is soul-destroying to exert a lot of energy to catch the sheep only to see them escape when a hurdle or gate falls over.

Keeping it Clean

If possible, have a source of water nearby so you can fill a footbath easily and wash away muck after handling. A fixed handling area that you use regularly will ideally need washing down to prevent spread of footrot and the build-up of muck. If washing is not an option then scraping and sweeping away muck and spreading some clean, dry material, such as straw or sawdust, can help to keep it clean for handling. In early summer, it is best to keep the sheep's fleeces clean. Dirty wool attracts flies, so turning sheep over in the muck or mud is a bad idea.

A perforated floor can be useful. The sheep stand on it in the pen or race (single-file walkway) and the muck drops through, keeping the feet clean before and after treatment. Sheep do produce a lot of muck and urine when they are collected straight from a field.

All this kind of infrastructure does cost a lot of money but if you're planning on having sheep on the holding for several years, it can save a lot of time and effort to invest in some decent handling facilities at the outset. Handling trailers can be useful if you are keeping sheep in different locations through the year. Agricultural suppliers have catalogues and websites full of fancy sheep handling pen ideas.

Working Under Cover

An existing barn, stable, lean-to or outbuilding can be converted into a sheep-handling area. In hot sunshine it can be a lot easier working in the shade and it is very unpleasant mucking about with sheep in the rain. If space allows under a roof, a small sick bay or quarantine pen next to the handling pens is a good addition. Any animals needing medical care can be easily separated from the flock and kept in the quarantine pen until you have finished treating them. Bear in mind that a sheep will never be happy when kept on its own in a pen, so always provide a companion. If a sheep does need to be kept inside, the animal should have access to hay and water at all times. It is a good idea to fit the quarantine pen with a hay rack and water bucket or plumbed-in water trough.

THE SHEEP RACE

A sheep race is a narrow alleyway through which the sheep walk in single file. It has a

A purpose-built sheep race is extremely useful for all manner of sheep husbandry tasks. It is made from galvanised sheet metal hurdles joined with metal rods, with a narrow gate at the entrance and a drafting gate at the exit.

gate at each end to control the flow of sheep. A race will often have a drafting gate at the end, which is useful when sorting sheep into different groups, for example when separating ewes from lambs. The race is useful for carrying out the following tasks:

- Drenching – administering a liquid medicine or mineral drench by mouth
- Injecting, where a vaccine or other injection is recommended in the neck
- Tagging adult animals or well-grown lambs
- Dagging – clipping the dirty wool from around the sheep's back end
- Checking teeth or tag numbers
- Weighing by running sheep individually into a weighing crate
- Footbathing by placing a long, narrow plastic or metal tray on the floor of the race; 7.5–10cm (3–4in) of water plus bactericidal salt or chemical solution. The race is used to make the sheep stand in the solution. The solid hurdles of the race prevent the sheep from turning around
- Examinations, for example checking teeth or treating eye conditions

HOW TO TURN A SHEEP OVER

Catching an individual will be much easier in a smaller pen rather than chasing a ewe round and round a larger space hanging onto her coat tails. Grabbing a sheep by the wool is unkind anyway, just like pulling somebody by the hair!

Once you have restrained it in a tight space, you may need to turn the sheep over to examine its feet. This can be done with two judo-type moves exercised simultaneously. Do not expect to master this immediately – it takes a lot of practice.

Stand to one side of the sheep, ensuring there is a sheep's width of free space behind you. With your forearm under the sheep's jaw or neck, lift it to momentarily take the weight off its front feet. As you do this, your other hand will be placed on the side of the sheep furthest away from you, holding the flap of skin adjoining the belly, just in front of the back leg. The hand and forearm under the sheep's chin lifts and turns the head away from you, and simultaneously the belly hand pulls up and backwards towards you, rolling the sheep off its back feet and depositing the animal on its side. As you roll the sheep's back end towards you, take a step backwards to keep your balance.

As soon as the sheep is off its back feet and on its side, you will need to hold the front legs, pull the sheep up and turn it so its back is facing your legs and its weight is on the thigh of one of its back legs. The sheep's shoulders will be against your thighs, with its legs pointing away from you. It's about as easy as learning to waltz by reading instructions, so if you have the opportunity to watch somebody experienced working with sheep, study how they handle them.

Sheep Handling Tips
Always catch the sheep under its chin first
If the shepherd has control of the sheep's head, this stops any forward movement and the animal will not struggle and cannot flee. This advice is the same whether working in a small pen or with the sheep in a race. Never lift a sheep by its wool as it will cause pain to the animal.

Keep the head up If the head dips down to the floor, the sheep can get to its feet and escape. When working with a sheep you have rolled over onto its back, always keep the head up off the floor. As soon as a sheep feels its head drop it will struggle. Keep the head up with your hands or knees at all times. If you have an assistant, ask them to support the head, neck and shoulders of the sheep while you do whatever it is you need to do.

Make the sheep comfortable Never hold the upended sheep upright so that its weight is on its tail. This hurts the spine of the sheep and it will struggle. The sheep's body weight should be on either the right or left thigh. If trimming the foot of the right back leg, have the body weight on that side, shifting it over when trimming the other back foot. Usually, if a sheep is struggling crazily while you hold it, it is because the beast is not comfortable. (Occasionally it will be an individual who gets in a crazy panic when handled and then it doesn't matter what you do – it will struggle.) The frontispiece of this book illustrates the position a sheep needs to be in for foot trimming.

Keep the shoulder off the floor A sheep can be placed comfortably lying flat on its side and often it won't struggle as long as you keep the toe of your boot under its shoulder and support the head. As soon as the shoulder touches the floor, the sheep can roll onto its feet again.

Watch the masters at work and study how they handle the sheep A good shearer uses their knees and feet to shift the weight and position of the sheep while their hands do the work. Study where the shearer's feet are when they are clipping the sheep's belly wool and the wool round the tail. This will give a clue as to the positions to adopt when trimming a sheep's feet.

TRANSPORTING SHEEP

If you need to transport sheep, then legally they must be able to walk in and walk out of a trailer purposely made for transporting sheep.

The ramp should not be a steeper angle than 45 degrees and should ideally have side gates to it. Check the condition and pressure of the tyres on the vehicle and trailer. Also check the trailer has a breakaway cable and a place to attach it to the vehicle. The towing capacity as stated in the vehicle manual should be checked, as well as the unladen weight of the trailer. The towing capacity is calculated according to the vehicle's braking ability. Essentially the important question is not 'Can the vehicle pull a trailer loaded with sheep?', but 'Can it stop in time with the weight of the trailer and animals behind?'

Sprinkle straw, sawdust, or wood shavings on the floor of the trailer to dry up urine and prevent the sheep from slipping over and getting mucky.

It is often a rite of passage for new sheep keepers and smallholders to take their maiden voyage of sheep transport in an unsuitable vehicle. I heard of a ram being transported in the back of an estate car. The ram suddenly burst through the back window, like an action hero, because he mistook his own reflection for a rival ram.

A small trailer is essential if you will need to move sheep on or off the holding by road. It can also be used for many more jobs around the smallholding.

Unloading weaned lambs from a large trailer onto fresh pasture.

CHAPTER 7

WEANING AND PREPARING FOR BREEDING

An annual breeding sheep sale of north country mules at Worcester market, UK. Every year, ewe lambs are brought from the northern hill country as weaned lambs in autumn and sold to farmers in the Midlands and south of England. The UK has a long-standing and sophisticated system of integration (stratification) between hill farms and lowland farms, making productive and sustainable use of land for food production.

For spring-lambing flocks, the sheep production cycle (or sheep calendar) ends in late summer with the weaning of lambs from the ewes and the sale of the first 'fat' lambs (new season lambs). A new sheep production cycle begins with readying the ewes again for breeding time in the autumn. In Britain the sheep calendar is as ingrained in the farming year as the sowing and harvesting of crops.

Autumn is a busy time in the traditional sheep calendar in Britain, as farmers sort out their flocks ready for the breeding season and the new production cycle ahead. Lambs are sold for meat before winter sets in. Ewe lambs are bought to join the breeding flock in future. Rams are purchased to bring new blood to the flock, and older rams and ewes no longer fit for breeding are sold for meat. There's a flurry of sheep movements up and down the country as the national flock is pruned into shape, ready for another productive year ahead. Sheep farmers and smallholders will be aiming to balance the number of sheep kept on the holding so that it matches the acreage of grass and free feed resource that will be available over the next twelve months.

WEANING

Weaning lambs means the separation of the lamb from the ewe. The lamb will no longer be able to suckle milk from the mother and has to rely on feeding itself by grazing grass. As there is no longer demand for milk, the ewes will naturally 'dry off' after weaning as their udders stop producing milk. The ewe's body re-absorbs the milk that is left and the udder shrinks back to the size it was at tupping time in the autumn. The ewe now has fewer physical demands on her and can use all the nutrition she gets from grazing grass to build up her own energy reserves ready for the next winter.

Ewes and lambs gathered in for weaning in July. In commercial flocks, lambs are weaned at around fourteen to sixteen weeks of age. For spring-lambing flocks this occurs sometime in July or early August. Ewes will be separated from the lambs and moved to a separate field. A good, efficient ewe will wean lambs whose combined weight is more than her own body weight.

Why Wean?

During pregnancy and lactation, the ewes will lose weight. Weaning the lambs from the ewes in mid- to late summer gives the ewes a much-needed recovery period before they are put back with the rams in the autumn. Weaning time also gives the shepherd a chance to assess the weight and growth of the lambs. Often the first batch of 'fat' lambs (weighing over 40kg/88lb), usually those raised as singles, can be sold shortly after weaning time. With good-quality pasture, the growth rate and the weight gain of lambs and ewes can be better after weaning, as the animals spend more time grazing and resting.

Without intervention, weaning would naturally occur for the lamb at around the age of six months. At that point, the ewe will try to get away from the lamb and prevent it from suckling.

When to Wean

In early-lambing flocks, where the lambs are creep fed (*see* box), weaning may take place as early as twelve weeks of age. On more extensive (outdoor/less intensive) sheep farms, weaning is carried out when the lambs are between fourteen and eighteen weeks of age. Weaning time is usually calculated according to the youngest lambs in the group. If lambs are weaned when the youngest are twelve to fourteen weeks old, the older lambs in the group will usually be sixteen to eighteen weeks. Just four weeks can make a noticeable difference to the size of the lamb by midsummer.

How to Wean

Gather the ewes and lambs into a pen and separate the lambs from the ewes. If possible, it is best to put the lambs into a separate field where they will be away from the ewes. For the first few days after weaning, the ewes and lambs will call for each other. If the lambs are

CREEP FEEDING

Creep feed is concentrate feed which is fed exclusively to lambs while they are with their mothers. This is done by putting the feed into a special creep feeder that has adjustable head or body spaces so only the smaller lambs can access the feed, preventing the ewes from guzzling it all. The lambs learn to nibble on the cereal or small pellets of sheep feed from a young age. By eight weeks, the lambs will be consuming a significant amount of extra protein and energy, increasing their growth rate. The advantage is that lambs can be sold at a higher weight and at a younger age, increasing returns to the farmer.

There are a number of disadvantages, however. Creep feeding incurs extra cost and requires additional labour. The meat quality also suffers – meat from 100 per cent grass-fed sheep is regarded as better and more nutritious. Creep feeding can also cause health problems in the sheep if the lamb's consumption is excessive. Rapidly growing lambs can be more prone to fatal clostridial diseases such as pulpy kidney.

Creep feeding is, however, a useful tool for lambs that are not thriving, either because they are orphans or the ewe is short of milk.

With larger flocks, a sheep race and draughting gate is used for separating the ewes and lambs into different groups at weaning time. This saves the exhaustion of having to manhandle every sheep.

out of sight and earshot of the ewes they will quieten down more quickly. Ideally, for the first two to three weeks after weaning, the ewes should be put on sparse grazing. This will ensure they do not get a flush of milk and will help them to dry off. Once dried off, the ewes can be moved onto better-quality grazing where the grass is more abundant in order that they may put on weight.

Weighing Lambs

Many sheep farmers will weigh the lambs at weaning time. Lambs weighing over 40kg (88lb) will be sold for meat or may be selected as future breeding females.

A benchmark of performance for ewes is comparing the weight of lamb that is weaned with the mother's own body weight. A healthy productive ewe will often rear the equivalent of her own body weight in lamb. For example, a ewe weighing 60kg (132lb) would have two lambs weighing 30kg (66lb) each at weaning time. If this is achieved in only sixteen weeks from a diet of grazed grass, it becomes a very productive and efficient way of producing meat for human

Weighing lambs using a weigh crate. The farmer uses weaning weights as a guide for assessing how well the management system and the sheep have performed, and for picking out lambs that are ready to sell.

consumption. These are the type of ewes that are most profitable to the sheep farmer.

There will be variation in the overall performance of the lambs each season according to the weather conditions, the pasture quality and the age and health status of the ewes, as well as the competence of the person looking after the sheep.

SHEPHERDING TASKS TO DO AT WEANING TIME

When the ewes and lambs are gathered in for weaning, it is a good time to give the sheep some health checks and treatments.

Health Checks of Ewes

Check the ewes are not lame by standing them for five to ten minutes in a footbath filled with water and zinc sulphate solution. Ewes that hold a foot up or limp after coming out of the footbath will need to be turned over (*see* Chapter 6) and the foot inspected and sprayed with antiseptic spray. If the same ewe has been treated for lameness on

two or three occasions in the previous year, consider adding her to the list of ewes not to breed from again or to sell as cull ewes.

Feel the ewe's udders. They should still be 'in milk' and soft and pliable, with no hard lumps or scar tissue from mastitis. If one side is without milk and there are signs of inflammation from a current infection or past mastitis, mark her with paint as one to cull or not breed from again, even if she seems healthy in herself. If the affected ewe seems lethargic or hot, or her udder is inflamed, take her temperature and consult a vet. Ewes' teats can also be damaged by lambs with orf infection (*see* below) so check for lesions and lumps on the teats that might hinder lambs suckling in the future. Wear gloves if you suspect orf might be a problem, as it is contagious.

Health Checks of Lambs

If there is lameness in the flock, weaning time is a good opportunity to footbath the lambs. Turn over and spray (with antibiotic or antiseptic foot spray) the feet of any lambs that are hobbling or holding a foot off the ground. If there is an infection deeper in the sole of the foot or between the digits of the hoof that cannot be reached by the spray then consider using an injectable antibiotic. See Chapter 15 for advice on treating and controlling lameness.

Check the skin around lambs' mouths, before and after weaning. Sometimes the stress of weaning in combination with bad weather or grazing too close to thistles can cause lesions and sores around the lamb's mouth. This is known as orf and is a skin disease caused by a virus. It is transmissible to humans and causes skin sores, so if you suspect any sheep may have orf, wear gloves when handling them.

Orf does usually clear up naturally after three to four weeks without treatment but sometimes a secondary bacterial infection can occur in the sores, so keep an eye on it. You can spray around the mouth with antibiotic spray if concerned. In the worst cases, an injectable antibiotic may be required. There is a vaccine available for orf, but it is a live vaccine so once it is introduced to the flock, it must be used every year. There's also a chance that the shepherd will give themselves orf when administering the vaccine to the lamb by inadvertently scratching themselves with the needle.

Wethers and Ram Lambs

Check the sex of the lambs as you sort through the flock and consider putting females into a separate group if you can. Wethers (castrated males) will be less distracted and should put on weight quicker if away from females. Check that wethers have been castrated properly. With inexperienced shepherds, the rubber rings used for castration of young lambs at lambing time are not always applied correctly. It's rare, but sometimes only one testicle will have been trapped by the rubber ring, leaving one to grow under the skin. If you find a lamb like this at weaning time, take it to an abattoir as soon as you can or seek veterinary advice on what to do, as it will be in discomfort.

Fully entire ram lambs should be separated from the female lambs and ewes by the age of about sixteen weeks and grown on in a separate group. Without attention to detail, an errant ram lamb could get ewes and female lambs pregnant before the 'official' tupping time has begun in autumn.

Ram lambs intended for show or sale will need feeding daily with cereals or concentrate if you are to achieve a good sale price.

Internal Parasites

After gathering the ewes and lambs, take fresh dung samples from about 10 to 20 per cent of

A Llanwenog ewe with her single-suckled ram lamb. Lambs reared as singles will grow at a faster rate than twins and often reach sale weight (more than 40kg/88lb) by weaning time.

the lambs and ewes and have them analysed for worm eggs. This can be done at home if you purchase an FEC pack (faecal egg-counting kit), or by booking in to deliver some samples to your vet. If the sheep have more than 250 eggs per gramme of faeces, they will need worming. Ewes will probably need a worm drench if they have been grazing where there were sheep the year before. Drenches can be bought over the counter from agricultural merchants, but it is important to consult a vet first as to which product to use, as on more than 90 per cent of holdings in the UK there is wormer resistance.

Feeding Ewes Post Weaning

Ideally ewes will be put back onto the same field they were on before weaning or onto a field where the grass is short after heavy grazing with horses or cattle. 'Rough' pasture, where there is more stalk and less leaf, would also be suitable. By restricting the quality and quantity of grazing for the first ten to fourteen days after weaning, the ewes' milk will dry up quicker and there is less chance of summer mastitis infections.

Once ewes have dried off, they should be moved onto better grazing. This allows them to begin putting on weight and condition, building up their strength before the tupping season comes around again in the autumn.

Feeding Lambs Post Weaning

In sheep flocks there may be a few lambs that are smaller and weaker than their peers at weaning time. This is usually because they did not receive adequate colostrum at birth or the mother has been short of milk (possibly because of mastitis). Separate these small lambs into a group if you can and feed daily with concentrate feed so they can catch up with the rest and stay healthy.

If dung tests show it is necessary, lambs can be wormed with a drench at weaning time and then moved onto 'cleaner' pasture (ground that has not had sheep on it for at least three months). Weaned lambs will grow rapidly if they have access to fresh new grass growth, particularly if it contains a mix of ryegrass and clover species and herbs like plantain and chicory.

The lamb on the right is an example of one that has not thrived in the field with the flock due to lack of maternal milk. It is narrower, with less muscle and fat than its peers.

CLEAN PASTURE

Grass which has not been grazed by sheep for at least twelve months is considered to be clean of the eggs and larvae of sheep gut parasites. This is difficult to achieve in practice as land type, lack of fencing and the amount of land available might not make it possible. Look at how you might be able to set up a rotational grazing system to benefit the health of the sheep.

STORE LAMBS

Store lambs are lambs that have been weaned but not yet reached ideal slaughter weight. Some farms specialise in buying store lambs at market and growing them on to slaughter weight and selling the meat to supermarkets. Mountain farms and farms in the hills do not usually have enough good-quality pasture for them to fatten lambs at home, so it is more economic for them to sell the lambs as stores to farms in the lowlands that have more fertile ground and better pasture.

AUTUMN STOCK TASKS

Assessing the Ewes' Condition

Late summer, before the start of a new sheep production cycle, is an important time to assess the ewes' condition before they are introduced to the ram. Ewes that are too thin after weaning often do not have enough milk when lambing time arrives again; thin ewes may also give birth to weak, underweight lambs.

Consider if a ewe will have enough 'time off' from work, between weaning and tupping (usually eight to ten weeks), to put sufficient weight back on to prepare her for the breeding season and winter ahead.

Older ewes that are known to be good mothers could have their lambs weaned earlier and be fed well after drying off to prepare them for winter. However, there comes a time when they will need to be retired from the breeding flock.

Consider retiring ewes whose spine feels prominent and the bones of the back and tail sharp with no fat cover. A thin ewe that is fed heavily during pregnancy to compensate will not put on enough weight herself to ensure adequate colostrum and milk production at lambing. The extra nutrition tends to go into the growing foetus, leaving her with a large lamb she cannot feed adequately.

A ewe in good condition before winter will have a light covering of fat and muscle; the spine will feel fairly smooth and the bones will not feel sharp in the middle of the back. Such a ewe is well prepared for winter and, with good-quality grazing, she should not need supplementary feeding in the first two or three months of pregnancy (unless there is frequent frost or snow).

A word of caution: after weaning, beware just selecting the fattest, best-looking ewes to purchase or breed from. The better performing ewes will be lighter because they have milked well for their lambs. Use lamb weaning weights as well as appearance to inform your decisions of how good a ewe might be.

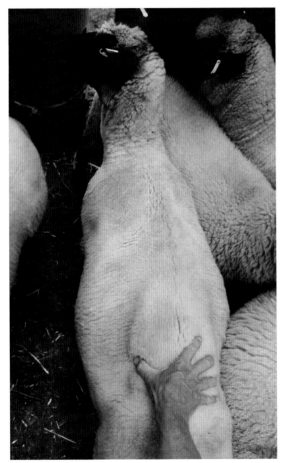

Assessing the ewe's body condition can be done by feeling the spine in the middle of the back. How much fat and muscle is the ewe carrying along her spine and round the tail and ribs? If the bones feel sharp under the skin, this indicates BCS 1, and this ewe is too thin to put to the ram.

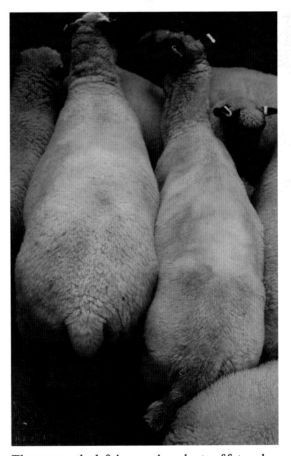

The ewe on the left is carrying plenty of fat and muscle and is in good body condition (BCS 3.5). A backbone too thickly covered with fat (BCS 5) would indicate an overweight sheep.

BODY CONDITION SCORE

The BCS is a simple 1–5 scale for judging what condition the sheep are in:

1. Very thin with no fat or muscle cover over the spine. Ribs, backbone and tail bones prominent.
2. A thin covering of muscle and some fat but the spinal bones and ribs still easy to feel. Overall lean condition.
3. A smoothing out of the spaces between the bones of the spine. The spine can be felt with light pressure. Good muscle layer and a nice fat layer over the tail bones. Overall good condition.
4. The bones of the spine and tail can only be felt with pressure applied. Muscle full, thick fat cover. Generally a bit overweight.
5. Spine impossible to feel. Very thick fat deposits around the tail. Sufficiently overweight to cause lambing problems or inhibit fertility.

Reasons for Poor Ewe Body Condition

Check for a reason why the ewe might be underweight:

Late weaning How many lambs has the ewe reared and what age were they weaned at? A ewe that has reared two well-grown lambs will have lost body condition from producing milk. Later weaning (after the lambs are sixteen weeks old) means the ewe will not have had time to recover body condition. Later weaning is not a problem if the ewe is in good strong condition (BCS 3 or above).

Age How old is the ewe? If she is over five or six with teeth missing, this could explain poor condition, especially in a system where the sheep are kept exclusively on grass and forage with no concentrates fed.

Parasites Has there been adequate internal parasite control in the flock in the period between weaning lambs and tupping? Consider analysing dung samples for worm eggs. The analysis should look for the presence of liver fluke as well if there are wet parts in the fields or there's a known history of fluke on the farm.

Lameness Has the ewe been chronically lame, preventing her from grazing?

Disease Chronically thin younger ewes may have an undiagnosed disease that is causing weight loss. Ask a vet if diagnosis is possible or cull poor performers.

When Should a Sheep be Retired from Breeding?

Deciding whether to put a ewe to the ram or not is an important decision. You also want the right ram to add to the gene pool of your flock and produce a new crop of lambs. Consider the following qualities of your ewes and rams when making your selection:

Lameness Does a ewe or ram have a recent history of lameness? It is not advisable to breed future breeding stock from an animal with poor feet; such an animal, moreover, is more likely to spread foot infections within the flock.

Udder health Has the ewe got a sound udder? A ewe that has had mastitis in the past or struggled to produce enough milk for her lambs should be retired from the breeding flock or sold as a cull ewe at market.

Temperament Has the ewe or ram got a quiet temperament? Occasionally an animal will be overly skittish or aggressive or very good at escaping. Sheep keeping is a challenging occupation or hobby so why make it harder by keeping animals like this? Rams that are aggressive towards humans should be culled on safety grounds.

Resistance to parasites Do certain individuals seem to suffer with frequently mucky bottoms and high worm egg counts? Genetically some sheep are more resistant to internal parasites than others, and it is a useful trait to select for. This can be done by culling poor-performing sheep and not breeding from their offspring.

Fertility If a ewe has run with a ram for six weeks and not got in lamb like the other ewes there is likely a fertility issue. Consider culling barren ewes as, without lambs to feed over summer, they will become too fat and less likely to get in lamb the next mating season.

Age The age to which a ewe can be productive in the breeding flock will vary. If the ewes are looked after well, they could breed six, seven or even eight crops of lambs, starting at two years old. That could be sixteen lambs from one ewe. Other ewes may only produce two or three sets of lambs before succumbing to problems like mastitis or lameness. In smaller flocks, the longevity of ewes tends to be better as they are cared for more closely and have less competition for food than in large flocks.

Checking Teeth

In a sheep market you may sometimes be surprised to see people climb into the sheep pens, hold a sheep by its head and look at its teeth. This is to assess the age of the sheep but also to check it has a good mouth. From the human perspective, a sheep's main purpose in life is to eat grass and produce milk, meat and wool. The mouth and teeth are very important for this aim.

There are variations between breeds, but most sheep will grow the first two adult teeth at the front of the mouth between twelve and sixteen months. Once the adult teeth have broken through the gum, the animal is no longer classed as a 'lamb' by abattoirs selling to supermarkets or butcher's shops.

During the second year of life, the young adult sheep will have two large teeth at the front and milk teeth either side. The animal will then grow two more adult teeth each year until the age of four to five, when it will have a set of eight. At this point it is referred to as a 'full mouth' ewe or ram.

Long in the Tooth

In sheep, the front teeth are only found in the bottom jaw, and they bite against a

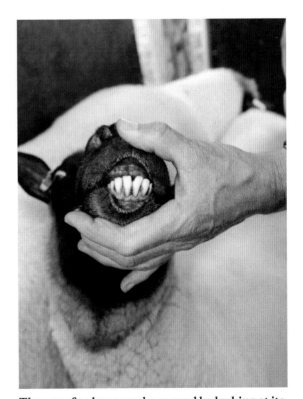

The age of a sheep can be gauged by looking at its teeth. This ewe has four large adult front teeth followed by milk teeth on the outside, indicating that she is three to four years old.

hard pad, or gum, on the top. Older sheep can be termed 'long in the tooth' when the front teeth grow too long and protrude in front of the top gum, thus impeding the sheep's ability to bite off the grass stalks. Occasionally, older sheep develop problems with their back teeth where the overgrown molars start to cause pain in the cheeks. Unfortunately, this is another reason for culling ewes, as sheep with mouth problems rapidly lose weight and are not strong enough to go through another winter and breeding cycle.

Broken-Mouthed Sheep

From about the age of six to eight years (it varies between breeds and individuals), most sheep will begin to lose their front teeth. When missing teeth, a sheep is called a 'broker' or 'broken-mouthed'. If broken-mouthed ewes are still in good body condition, they can be retained for breeding. Completely toothless old ewes with no front teeth tend to fair better than ewes that have three or four 'pegs' left at the front. With good, plentiful grazing, the toothless old sheep can still pluck the grass away with their gums. If being retained over winter, they will however require good-quality forage as hay or silage and some concentrate feed to support their body condition.

What to do with Old Sheep?

Deciding on what happens to old breeding ewes will depend on your reasons for keeping sheep and how much grass you have available. If productivity is important for self-sufficiency or commercial reasons, selling older ewes at market is the best option. There is a strong trade in the UK in older sheep and they are usually sold through livestock markets. Buyers from abattoirs attend to buy the 'cull' ewes and rams for mutton. The meat from older sheep is an important supply for ready meals, fast

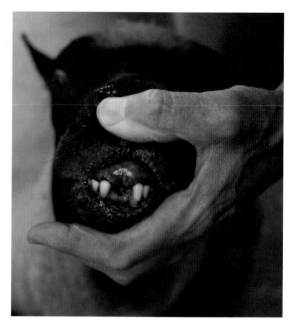

A broken-mouthed ewe. As ewes age, they lose their front teeth, which hampers their ability to graze. Without supplementary feed, they lose body condition and may struggle to produce enough milk for their lambs.

food takeaways, restaurants and religious festivals. Prices vary from year to year and season to season, but expect to get about a third to a half of the original price paid for a yearling breeding ewe.

On a limited acreage, tough decisions need to be made about who stays and who goes. Retired breeding ewes will eat grass that could feed ewes with lambs at foot. Keeping fewer productive breeding ewes will mean fewer lambs and less income. There may also be expenses incurred on the old ewes through veterinary, medicine and shearing bills with nothing coming back in return.

Pets or Lawnmowers

If income from the sheep is not a priority, retired ewes can be kept for the purpose of keeping the grass down and providing you with company. You could see it as an eco-friendly four-legged lawnmower that fertilises the soil as it goes. Remember you will still need to provide all the daily welfare checks plus health checks through the seasons – foot treatment, dagging, shearing, protection from blowfly, checking for worms and liver fluke, providing forage and feed through winter if necessary. Animal welfare should always be top of the to-do list!

Euthanasia

If you decide not to sell your older ewes at market but to let them die at home on the farm, there will come a time when a decision must be made about when to ask a vet to put an old or sick sheep to sleep. A vet will do this if they think it will prevent prolonged suffering of the animal in the event of severe illness or old age. It is important that a sheep is not left to experience unnecessary prolonged suffering, especially if it can no longer stand up on its own and eat and drink.

Disposing of Fallen Stock

Since 2013, it has been illegal in the UK to burn or bury the carcasses of farm animals on the farm or smallholding. Collection of the carcass by a licensed fallen stock collection company needs to be arranged and paid for. These can be found on the National Fallen Stock Co. website (nfsco.co.uk).

Selecting Replacement Breeding Stock

Ewe Lambs or Yearlings?

New breeding females will be required to replace any older ewes that have been retired, sold or died on the farm. The easiest option is to select the best of your own home-bred ewe lambs to keep. Alternatively, ewe lambs of any breed can be bought in one of the early autumn breeding sheep sales or purchased privately from another farm. They are usually kept for one calendar year before they are run with a ram.

Buying yearlings (eighteen-month-old ewes that have not bred before) is an alternative to ewe lambs. Yearlings are more expensive to purchase - usually at least 30 per cent more –but could produce lambs within six months of arriving on your farm, as they are ready to go to the ram straight away. Prices vary according to breed, quality and market conditions.

SHEARLINGS AND HOGGETS

Year-old sheep are also referred to as shearlings, because they have been sheared once. Two-shear or three-shear ewes simply mean two or three year old ewes.

Ewe lambs are also known as hoggets or hoggs – a hogget can be defined as any sheep between weaning and first shearing.

Lambs under twelve months of age can be put to the ram but this is only advisable if they are well grown and over about 70 per cent of their finished mature body weight. Consider carefully if you need to breed from ewe lambs as they are more likely to need assistance at lambing time. Hoggets can be short of milk and don't always bond with their lambs immediately.

Suitable types of ewe lambs to breed from in their first year include lambs from vigorous cross-breeds, such as Welsh or north country mules, and lambs born in January or February and put to the ram in late October or November. The latter can be suitable to breed from as they will be over twelve months of age by the time they give birth. Careful attention needs to be given to the feeding of breeding ewe lambs throughout pregnancy, lactation and after weaning to ensure they will go on to be healthy productive ewes in the future.

Selecting ewe lambs to keep for breeding out of a group of seven-month-old lambs. Selection criteria for a good ewe lamb include: the milkiness of the dam, good weaning weight of the lamb, feet in good condition, clean back end, good-quality fleece, correct mouth, generally bright and alert and no defects or asymmetry in the way it stands and walks.

Ewe lambs giving birth at twelve months of age will struggle to raise twins due to lack of milk. The usual practice is to take one lamb away and bottle rear it or foster onto another ewe, but this also creates extra work at lambing time. Be careful not to overfeed ewe lambs carrying single lambs to avoid difficult lambings.

Points to Consider when Selecting Replacement Stock
Size Larger lambs with a faster growth rate should be evidence that they have mothers with plenty of milk and this trait will hopefully be passed on to their offspring. Be aware that single-born lambs will have received more milk than lambs raised with a twin. Although singles tend to be larger, they don't necessarily make the best breeders.

In a flock where the aim is to keep productivity high by producing as many lambs per ewe as possible, it is common practice to keep ewe lambs that are twins themselves. These will not be as large as the lambs raised as singles, but they will be more likely to produce twins when they breed.

Weaning weight Weighing lambs at weaning time can give a good indication of the mothering ability and milkiness of the ewe. Assuming the lambs are weaned from the mother at around sixteen weeks of age, a good ewe with twin lambs will wean more than her own body weight in lamb. If this is achieved chiefly from a diet of grass, the ewe will be a productive, efficient and profitable animal to keep.

Closeness in age Selecting lambs of a relatively even size and age can be an advantage when it comes to feeding them as a group. Lambs of a similar age (born within four weeks of each other) will be ready to breed with a ram at the same time.

Feet Do not keep lambs to breed from if they are out of ewes that have poor foot shape or regularly suffer from lameness. It is important for the sheep to have a tidy foot shape with no cavities where the wall of the hoof meets the sole of the foot. Reject any with obviously overgrown feet or lambs that stand unevenly or that roll their feet to the side when they walk. Watching sheep walk on concrete or similar hard floor will give you the best view. Sheep that stand 'up on their toes' are usually favoured by shepherds over those that walk with weak pasterns and slouch down on their heels. If interested in purchasing a pen of sheep at a market, try to see them walking along the alleyway, not just standing in the pen. You could even ask the vendor if you can see them walking before the sale.

Teeth Inspect the mouths of any sheep that are earmarked as replacements. The front teeth in the bottom jaw should meet the pad of the top jaw evenly, being neither excessively forward nor back. Be mindful that the teeth will grow as the sheep matures and if too far forward to begin with, may over shoot the pad at the top, impeding the sheep's ability to eat. Similarly, reject any individuals that have 'parrot jaw', where the bottom jaw is not long enough and the teeth are set too far back in the mouth.

Eyes Look for animals whose eyes are bright and alert. Paleness or yellowing of the membranes inside the eyelids can indicate anaemia, possibly caused by liver fluke infection.

Udders Ideally a ewe will have a tidy udder of a good shape. When the udder is full of milk, the teats should not be oversized or too close to the floor. This can be a problem in older ewes that makes it difficult for newborn lambs to latch on and suckle. In ewes that have raised lambs before, check for signs of mastitis and damage to the teats. You will not be able to assess the conformation (shape) of sheep's udders if they have not yet bred but if you know that the mother of a ewe lamb you are thinking of keeping has had trouble then consider if you should breed from her daughter.

Wool quality and breed points When choosing flock replacements, the quality of the wool may be a consideration. The fineness of the fibre, length, colour and appearance are all important considerations when assessing wool. Wool quality can also give clues to the health of an animal. A healthy animal will have a thick, bouncy fleece with some lustre to its appearance.

Pedigree breeding If the intention is to breed pedigree sheep of a specific breed, pay attention to the breed points as set out by the specific breed society. These are characteristics in a sheep that the breed standards set out as ideal. For example, in the Llanwenog breed, two desired characteristics are short ears set at an angle of ten minutes to two o'clock, and no brown wool around the head or tail.

Spend time looking at sheep in fields, agricultural shows and markets. The more you look at them, the more you will get your eye in and be able to compare the good and bad points.

CHAPTER 8

ALL ABOUT RAMS

A Llanwenog ram in full fleece in autumn, Ceredigion, Wales.

HANDLING RAMS

Safety

A ram can seem as quiet as a lamb when with the flock but beware of handling one when he is on his own. A lone ram can be dangerous unless he is pressed into a tight space with gates or hurdles. Do not give a ram enough space to back away and lunge at you. Rams use the blunt force of their heads to head butt. Once the ram is in a tight space you can grab him under the chin and hold his head up. Even with his head secured, be aware he may jump upwards unexpectedly so keep your chin out of the way. Never put your head near an unsecured ram. Rams are more likely to put their head down and butt if they feel threatened. Dogs and small children can put a ram on alert. Do not approach a ram head on or try to catch him when he is in a large pen or paddock by himself.

Don't be lulled into a false sense of security if a ram does become tame. A tame ram can sometimes be more dangerous than one that is not used to being fed or following a bucket, as they get playful and swing their heads about at feeding time. A tame ram might try to mount you when you bend over to clean some leaves out of the trough or he may mistake you for a brother and give you a playful head butt. Keep a respectful distanced formality to your interactions but be firm and confident when restraining him.

Catching Hold of Rams

Restrict the ram's space and keep him moving along with other sheep, and he will be fine. If you do need to catch a lone ram, first try to lure him into a small pen or sheep race with some food. If he is not bucket trained, move him into a small pen from a larger one.

If possible, bring a ram in with some other sheep. A group of rams is easier to catch than one on its own. If you can get the ram pressed into a pen with other sheep and no spare space, you can usually catch hold of him under the chin quite easily.

If the ram is on his own in a paddock, the best way to catch him will depend on whether he is bucket trained or not. If he is bucket trained, he may follow a bucket into a small pen to eat. If the ram is not tame and he is on his own, he will go into flight mode when you try to move him. This could mean he will run away from you and be impossible to chase through the gate or he could pose a threat and adopt an aggressive stance ready to butt you or the sheepdog (if you have one).

In this situation, it is better to take a bit more time to plan how you catch him. If several days are available, you could start feeding him and gradually taming him. Alternatively, you could fetch some other sheep and release them into the paddock with him, then gather them all into a pen or trailer as a group. He will usually come quietly when in female company. If he is a docile animal, you may be able to usher him along into a pen on his own.

Turning Over a Ram

The same technique for turning a ewe over can be used on a ram but it will usually take two people. A ram will typically weigh around 30–40kg (66–88lb) more than a ewe. That could mean a hefty 80–120kg (176–265lb) you are trying to flip over.

Flipping a ram is achieved more by speed and technique than brawn. Experienced professional shearers and sheep farmers are often small and wiry, but the best will happily flip a ram weighing twice their own body weight onto the shearing mat ready to clip his belly.

If you are not quite ready for this but you need to turn a ram over, having an assistant can help when pulling off the following move.

The first person holds the ram under the chin, restraining him from moving forwards. Upward pressure is applied from the forearm under the ram's chin to take the weight off the ram's front feet. While this is happening, the second person, standing alongside the

A ram lying on his side on the shearing board.

If possible, handle a ram in a group with other sheep.

ram, grabs the ram's furthest back foot and pulls it under the ram towards the ram's shoulder on the opposite side. This should sit him down onto his rear end. The first person, holding the neck and shoulders, can then roll him onto his back, grabbing him by the upper front legs ready for foot treatment or shearing.

CONFIDENT HANDLING

It is no good trying to handle a ram if you are too nervous. If you are frightened, that will make two of you, and it will lead to a nervous, unpredictable encounter. As discussed earlier, if you have a suitable solid pen set-up, you can handle the ram or rams confidently. In whatever move you are trying to pull on the ram, always remember to keep your head and face away from his own head and feet in case of an unexpected head jerk or kick. Hold him by the head or neck but do not hug him tight.

Halter Training a Ram or Ewe

Bring a young ram or ewe into a pen regularly and the animal can be trained to wear a halter. This is a piece of rope that loops around his muzzle and the back of his head. Try putting a halter on the sheep a few times a week and leave it on for approximately twenty to thirty minutes. The animal will soon be accustomed to it without panicking. When it is calm wearing it, it can be left tied up for a short while. Once this is achieved, you will be able to walk along while holding the rope. You will need to get to this stage if planning to exhibit the sheep at an agricultural show or sell it in a pedigree sale.

HOW MANY RAMS?

For flocks under fifty ewes, one ram is usually sufficient to get all the ewes pregnant within a six-week time frame, although you might decide to keep two rams to vary the gene pool and split the flock into two groups at tupping time. The rest of the year, when they are separate from the flock, the rams can be pals.

THE BREEDING SEASON

For most sheep breeds, the mating season begins in late summer and extends into late autumn. The date the ram is introduced to the ewes will dictate the date when the flock begins lambing in the spring. The average gestation length for sheep is 147 days or approximately five months (*see* Chapter 12).

Traditionally, lowland flocks in the warmer south of Britain will be the first to start lambing in the new year. For these flocks, the rams will be turned out with the ewes in August. Flocks in the north and uplands will follow by turning out rams in the autumn.

The ram knows when a ewe is ready for breeding, and a ewe will only 'stand' for a ram when she is ovulating. Ewes begin coming into season (ovulating) when the day length becomes shorter. Later tupping, in mid-October to November, leads to a tighter lambing period, as all the ewes should be in an ovulation cycle by then.

Apart from the Dorset breed of sheep and some breeds with a Dorset influence, most ewes only naturally begin to come into season and release eggs when the days begin to get noticeably shorter in September.

RAM HEALTH CHECK

Research has shown that the sperm that will fertilise the eggs is made in the ram's testes seven weeks before mating. Because of this, attention should be paid to the health of the breeding ram two to three months before he is needed. For March-lambing flocks, where the ram is turned out at the start of October, this would mean giving the rams a health check at the end of July. Make sure any foot infections are dealt with promptly, check the teeth and the testicles for any abnormalities and treat for internal parasites if necessary. A ram that has a temperature due to infection can have reduced fertility so it is worth keeping a closer eye on the ram's health and vitality in the three months before tupping begins.

WHAT AGE OF RAM TO CHOOSE FOR BREEDING

Ram Lambs
It is possible to breed from ram lambs between six and twelve months of age. A ram lamb can be run with a small group of ewes (fewer than twenty) to test out whether he is fertile. Usual practice would be to leave a ram lamb with the ewes for two to three weeks and then take him away and put a tried and tested older ram with the same group of sheep. That way the shepherd can be certain all the ewes will be in lamb after six weeks. Buying a ram lamb can be more cost-effective than buying a yearling ram and the ram can

be tested a year before he becomes the main stock ram.

You might decide to keep a nice ram lamb and sell him as a yearling. Using him on some unrelated ewes as a lamb will keep some of his genes in the flock and give confidence to the buyer the next year as he will be proven to be fertile. The size of a ram lamb matters. If used for breeding at under twelve months of age, he should weigh approximately two-thirds of his estimated mature body weight when turned out with the ewes.

Yearling Rams
A fully grown young ram is a handsome sight, and yearlings will command the best prices in breeding sheep sales. Yearling rams are usually between eighteen and twenty months old when they are sold, have been sheared once and will have the first two adult teeth at the front of the mouth. You may see some redness around the gum of a front tooth when inspecting the mouth, but this is normal if the adult teeth are still coming through. Ask the breeder if they have been used to breed before.

Avoid rams that are overfed. Some breeders feed their pedigree rams a lot of

Well-grown Llanwenog ram lambs. These could be used for breeding in their first year.

concentrate feed from when they are lambs and this can negatively affect their fertility and longevity. A ram raised predominantly on grass will not look as impressive but should be in better health than an intensively reared specimen.

Senior Stock Rams

Buying an older ram can be a viable option for small flocks. A well-looked-after ram can be used to breed up to around the age of eight, maybe more in some breeds. A senior stock ram that has been used in a flock for three or four years will be a proven stock getter (fertile) and likely of sound feet and good temperament, so could be a good purchase and often at a lower price than a yearling ram.

CONSERVATION BREEDING WITH RARE BREEDS

You may decide to link your sheep enterprise with the conservation of a rare UK breed. If this is the case, ram selection and purchase may be different because of a smaller available pool of genetics. Breed societies are important in helping buyers choose the right rams to use on their rare breed flocks. Some rare breeds are still very focused around their place of origin. For example, the Llanwenog breed still has a strong core of breeders in west Wales where the breed originated. The Lonk is a hill breed from the Lancashire Pennines in the north of England. Both these breeds have their main annual sales in the region where the breeds were created.

Annual Llanwenog breed society show and sale. Gareth Lloyd and family with his prize-winning ram.

A Lonk ram for sale in the annual breed society sale.

SELECTING AND BUYING A RAM

If a ram is to be used to produce female flock replacements then selecting a good ram is an important task. His genes will have a large influence on the flock for years to come. Defects, such as poor feet or bad temperament, are often passed on to the next generation.

Rams can be bought privately from another farm or from a pedigree or commercial auction sale at a market. Think of your breeding objectives beforehand and keep a budget in mind.

Prices for pedigree rams will vary according to the popularity of the breed and the number of sheep available. Looking at past sale reports on auctioneers' websites and online adverts can give a guide to values. In pedigree breed sales, there are usually one or two sought-after rams that will fetch a high price, usually from well-established breeders, and the rest of the rams on offer will be sold for significantly less.

Rams for commercial cross-breeding will be sold through auction markets from August to October in the UK. These are usually from terminal sire breeds like Texel, Suffolk, Charollais and Beltex. Plenty of these are usually available for a reasonable price. Use auction mart reports, pedigree sale reports and private online adverts to determine values of different types of ram. The meaty terminal sire breeds will have the best value when selling for meat at market as a cull ram when you no longer need them as a breeding animal.

TERMINAL SIRES

Terminal sires are sheep that are used primarily in commercial meat production. They are so called because their offspring are sold for meat and not retained in the flock for future breeding. They will be carefully cross-bred with ewes whose genetic qualities complement theirs to produce good-quality lambs for market.

This Roussin ram has a good set of teeth and bright eyes with no signs of anaemia (yellowing of the white parts or membranes around the eye). He has been used for three seasons as a terminal sire, running with a group of forty ewes for six weeks.

Points to Look for When Selecting a Ram

Whether it's for commercial sheep meat production, wool production, pedigree sheep showing or rare breeds conservation, the type of ram you choose will depend on the breeding objectives you set for your sheep flock and the planned end product. Decide on what kind of sheep enterprise you wish to run before buying your breeding stock.

As with purchasing ewes, the starting point should be to select a healthy-looking animal that stands and walks well and has no obvious defects in feet, teeth and general appearance. A healthy ram will be in good body condition but not over-fat. A ram should have bright, alert eyes, no discharge from the eyes and nose, will spread its weight evenly on all four feet and will have no sign of scouring (diarrhoea).

The Ram's Testicles

When it comes to rams' testicles, size matters. Research has shown that scrotal circumference is a good indicator of fertility and the amount of sperm the ram will produce.

The ram's testicles are obviously the important appendage required for breeding. If possible, before purchasing a ram, do a physical examination. Have one person holding the ram's head still in a narrow space where he can't turn around. The person examining runs their hands from above the testicles downwards over each one. They should be close to an even size with no lumps, shrivelling or obvious defects. At the bottom of each testicle is the epididymis, where the sperm is collected; it feels like a smallish extra ball that is attached to the main rugby ball-shaped testicle above. Rams can suffer from an infection here, so check both are a similar size and free from signs of inflammation. Reject any ram where one testicle is significantly larger than the other or where signs of inflammation or scar tissue can be felt.

Appearance and Breed Points of Interest

Appearance is everything in the pedigree breeding and sheep showing world. If buying a ram for pedigree breeding, refer to the relevant breed society for a list of the desired breed points.

Cross-Breeding with a Terminal Sire

If having healthy, vigorous lambs for commercial meat production is your main aim then consider cross-breeding your ewes with a ram of a different breed. Cross-bred offspring benefit from a factor known commonly as 'hybrid vigour' and the lambs will usually outgrow their pure-bred counterparts. Cross-breeding is often used to improve the carcass quality of the lambs. This is done by using a ram from one of the terminal sire breeds on your native-bred ewes.

For example, a Charollais ram, originating from France, might be used on an improved (larger type) Welsh Mountain ewe, or a Roussin ram (another French breed) on Llanwenog ewes (a west Wales breed). In years gone by, the Suffolk was the most popular breed for producing cross-bred butcher's lambs, but these days the terminal sire used most commonly in Britain is the Texel.

The use of a terminal sire gives improved vigour, growth rate and meat yield in the lambs but can present more problems at lambing time and make harder work for the ewe if she has two very demanding, fast-growing offspring to feed.

QUARANTINE AND SETTLING IN

Aim to purchase a ram and have him on your farm well in advance of the time when you will be turning him out with the ewes. Ideally, a new ram will be placed in a separate field away from the flock to quarantine for a period of three weeks. This way you can ensure that he is free from footrot, sheep scab or other health problems. By bringing him onto the holding early, you can let him settle in and you will have time to remedy any problems before he is needed to run with the ewes. It is not unknown for a ram to drop dead the day after he has arrived on the farm or, worse, the day before he is due to go out with the ewes!

TEASER RAMS

A 'teaser' is a ram that has been vasectomised by a vet. He still has all his working apparatus and ram hormones but is essentially infertile. Teaser rams are used on sheep farms to help synchronise the time at which the ewes come into season and ovulate. The teaser ram is usually run with the ewes ten to fourteen days before the date at which the shepherd wishes breeding to begin for real. The aim of this practice is to shorten the lambing period in the spring so the shepherd has a busy but shorter lambing period and the lambs are born close together. This also gives the advantage that as the lambs grow they will be a similar size and easier to manage as a group later in the season. A teaser ram can be kept year after year and used to run with the ewes until he is quite an old fellow.

CHANGING THE STOCK RAM

The stock ram is the main breeding ram in the flock, of proven quality and over two years old. If retaining ewe lambs each year to grow the size of the flock or provide flock replacements, you will need to sell or retire the stock ram and buy another unrelated ram to breed with the daughters of the stock ram to avoid birth defects and abnormalities caused by in-breeding. For future reference, keep breeding records of which ram individual ewes have run with and the parents of each lamb.

TUPPING

Raddle Marking

A coloured paste, known as 'raddle' is usually brushed onto the ram's chest, between his front legs, before he is introduced to the ewes. Its purpose is to show when a ram has mated with a ewe, as the colour is left on the ewe's rump when she has been served by the ram.

The paste is a combination of coloured powder (bought from agricultural merchants) with cooking oil, mixed to the consistency of thick cream. The raddle is applied with a flat stick, wooden spatula or a paintbrush. Two people are usually required for this, as the ram is turned over onto his back to apply the paste, although it is also possible for the raddle to be painted on with the ram standing up. In this case, the animal's head must be well secured with a rope halter for safety.

A note can be made in the diary of how many ewes have been mated each day, so the shepherd can get an idea of when the ewes will be lambing. Some shepherds change the raddle colour after fourteen or twenty-one days to give an indication of which ewes will be early lambing and which later. Raddle marking ewes also tells the shepherd if the ram is working or not. Occasionally a ram can be ill, lame in a foot or just refuse to mate with the ewes. If this is the case, it will need to be changed for another ram.

Sweeper Ram
A sweeper ram is used to try to ensure all the ewes get in lamb when a young, unproven ram is being used. Unless fertility tested by a vet, the shepherd cannot guarantee a

A group of rams. If you rely on producing lambs for your income, it is important that as many ewes as possible become pregnant. It can be useful to have a ram or two in reserve in case the preferred ram has a problem during the critical breeding period.

young ram is fertile, so, to hedge their bets, the shepherd will introduce a ram that is a proven stock getter. Usually the unproven ram will be given the first fourteen or twenty-one days with the ewes and will then be replaced by the sweeper ram.

HOW LONG SHOULD A RAM BE WITH THE EWES?

Traditional practice on UK sheep farms is to leave the ram with the ewes for six to eight weeks. The ewes' oestrus cycle (when eggs are released by the ovaries ready to be fertilised) is on average seventeen days, with a range between fourteen and nineteen days. A ewe's oestrus is considered to be when a ewe will be receptive to the ram, which usually lasts for twenty-four to thirty-six hours. Leaving a ram with the ewes for six weeks gives each ewe at least two full oestrus cycles and a good chance of becoming pregnant. Taking a ram away on a set date will give a clear cut-off date for when lambing will end in the spring.

A flock of ewes that have been served by different rams. If it is necessary to record the sire of each set of lambs born, a different raddle colour can be used for each ram and a record made.

KEEPING A RAM OUTSIDE THE BREEDING SEASON

If having a lambing cut-off date does not concern you, the ram can be left with the ewes all winter, although you will have to decide if it is good for the ewes to have a ram competing for their food. A boisterous ram pushing in for food can also be a danger to the person feeding the sheep. A ram can be kept with other species of grazing animal for company but would prefer to have another sheep to graze with. One or two castrated male sheep (wethers) could be useful for keeping the ram company and could be eaten or sold as mutton later if so desired. If more than one ram is kept on the farm, the rams can be kept as a small group away from the rest of the flock between January and the start of the next breeding season later in the year.

Ram lambs fighting. Rams will fight when first introduced to one another but usually settle down after a few hours.

CHAPTER 9

SHEEP HOUSING

Sheep housing makes it easier for the sheep and the shepherd during harsh winter weather, particularly for pregnant ewes carrying more than one lamb. Lamb losses in the spring can be higher if ewes carrying twins are lambed outside.

PRINCIPLES FOR HEALTHY SHEEP HOUSING

- Provide cover from the rain but maintain good airflow and ventilation without being draughty at floor level. The importance of good ventilation for the health of housed sheep cannot be overstated.
- Provide a clean and dry lying area with suitable bedding, such as straw, old hay or woodchip. Wheat, barley or oat straw is the most effective and comfortable bedding for sheep housing. Try to buy small bales from a local farm or hay and straw merchant.
- Make sure there's enough fresh bedding, top up daily as required to keep it clean and dry, particularly during lambing. Straw is the favoured material for most sheep farms as it is absorbent of moisture and easy to spread. Small bales are the most convenient, unless you have a tractor with a loader, you cannot move the large bales.
- Minimum (legal) lying space for pregnant sheep in the UK is 1.2-1.4sq m (13-15sq ft) per ewe rising to 2-2.2sq m (22-24sq ft) if housing the ewes with lambs at foot. Minimum lying space for pregnant ewes in organic/higher welfare systems is 1.5-2sq m (16-22sq ft). In practice you will probably want to allow more space than this.
- In addition to lying space, the pregnant ewes should have at least 2sq m (22sq ft) per head for exercise and feeding space.
- To minimise bullying, the housing should have enough feed space so that all

To ensure high welfare, housed ewes should have adequate feeding, lying and exercise space and ample straw bedding.

sheep in the pen can eat at the same time without competition and aggression.

- Ensure ewes have 45cm (18in) of trough or feeder space per ewe when feeding concentrate and forage. Without adequate space, ewes will jostle and bump each other to get at the feed. In doing so they may get injured and the shy ewes will not receive adequate nutrition.

FEEDERS

Sheep will not eat wet or dirty feed, so when feeding sheep, particularly outdoors, a specialised sheep feeder is a must-have. Throwing hay on the floor is extremely wasteful, as the sheep will trample most of it into the mud underfoot. Sheep will not eat hay that is wet, so a covered feeder is useful for ad lib feeding outdoors.

Never use a horse haynet for feeding sheep, as they are likely to become tangled in it. There are many different types of sheep feeder on the market.

Simple Trough

This is essentially a tray that goes on the floor, and is suitable for feeding concentrates or cereals. It keeps the feed free of dung or mud. Wipe out with a handful of hay or straw before feeding the sheep. Turn the trough over after the sheep have eaten to keep it free from mud and dung before the next feed. Do not use it for feeding hay.

Creep Feeder

This is a covered feeder where concentrate feed is left for lambs to nibble on to supplement their diet. The feeder has adjustable bars so mature sheep with larger heads cannot access the feed. Care needs to be taken, as fast-growing lambs fed concentrate can be more prone to metabolic disease or sudden death from clostridial disease. This is also unsuitable for feeding hay.

Covered Hay Rack on Wheels

This is a very useful and versatile type of feeder that can be used for feeding hay as well as concentrate. It's handy for feeding ad lib hay (meaning the sheep have access to forage all the time), as the roof keeps the forage dry. A tray underneath the central hay rack can be used for concentrate feed. The wheels allow the shepherd to move the hayrack regularly, preventing the ground getting too muddy around the feeder. This kind of feeder can also be used inside a barn at lambing time.

For small flocks, a covered feed trough and hay rack on wheels is very versatile. It can be used indoors or outdoors for feeding hay and concentrate, can be moved easily and has a lid to keep hay dry in the field.

Ring Feeder

This is used by farmers to feed big round bales of haylage or silage. The advantage of big bales is that a tractor can be used to take a large quantity of forage to the sheep, but the disadvantages are the cost of equipment, and that in wet weather the tractor will cause damage to the soil and ground compaction. To make big bales practical, the flock should be larger than twenty-five sheep, so the forage gets eaten before going off. Ring feeders will not keep hay or haylage in good condition in wet weather but they can be used effectively inside a barn when sheep are housed.

Fixed Hay Rack

These are usually made out of metal welded mesh with 5cm (2in) holes. The old antique ones are beautiful and made with closely spaced vertical wooden bars. A hay rack fixed to a wall or internal fence is useful for feeding hay inside a barn where the feeding location will not change.

Walk-Through Feeder

Consisting of a wide trough about 80cm (31in) wide with a 'feed fence' of two or three metal or wooden rails each side, a walk-through feeder is useful in lambing sheds as two groups of sheep can be fed simultaneously. The shepherd can feed the

A ring feeder is useful for feeding large round bales of hay or haylage, and can be used inside a barn or outside.

sheep without being bumped and jostled by hungry ewes. A walk-through feeder is suitable for feeding hay, silage, haylage and concentrate feed. The disadvantages are they are bulky and heavy to move around, and take up quite a lot of internal floor space.

TYPES OF FARM BUILDINGS SUITABLE FOR SHEEP

Pole barn

Such barns are often constructed from old telegraph poles with timber purlins supporting a tin roof. Yorkshire boarding (timber planks with a gap in between each board) makes a good ventilated but sheltered space for sheep. Check the base of the poles are still solid and not rotted through.

Polytunnel

Many farms and smallholdings use polytunnels for lambing. Ventilated mesh sides are required up to a metre (3ft) high on the sides, and one end of the tunnel should be left open or with a ventilated mesh door. If early lambing, in January to February, you could muck out the polytunnel after lambing and still grow a crop of vegetables.

Stable

Many stable buildings are poorly ventilated at roof level. A stable can be used for a hospital pen or handling pen but not for prolonged housing periods unless the ventilation above the sheep is good. Leaving the stable door open is not enough – there needs to be somewhere for warm, moist air to escape at roof level.

Traditional Stone or Brick Barn

The number of sheep inside a building will determine how much ventilation is required. Fewer than around eight sheep in a large internal space shouldn't be cause for too

A traditional stone barn used for sheep housing. It has 'arrow slits' in the stonework for ventilation. An open doorway and a vent high up on the back wall allow warm, moist air to escape.

much concern in a building that is open internally to high eaves. More than a dozen sheep, however, will create a considerable amount of ammonia and moisture-laden air, so it could be an advantage to improve the airflow if possible, or only bring the sheep inside at night.

If you have taken on an existing farm or smallholding, it is likely there will already be buildings that can be adapted to make sheep housing. Some old barns will have ventilation (arrow slit) holes in the walls that can be reopened. Assess if there is adequate airflow through the building without a whistling draught at floor level. Remember, the more sheep that are housed, the more heat and moist, ammonia-laden air there will be needing to escape. Hanging cobwebs indicate that ventilation is poor. Letting off a smoke bomb device can indicate if there is a sufficient dispersal of stale air from above the sheep pens and a flow of fresh air into the space.

Field Shelter

Field shelters can be useful for catching and handling sheep or temporarily housing a small number for short periods, for example if a ewe is ill. They are not so good when being used for a longer period, as the ground will usually become very muddy and slippery around the shelter. The shepherd will also have to trudge across the field to check the sheep in the shelter, which is less convenient than having the sheep in a barn on the yard with electric lights installed for nighttime checks.

Modern Portal-Frame Barn

Steel-framed barns are favoured on modern farms because a larger area can be spanned with a roof without intermediate supports breaking up the floor area inside the barn. This makes the barn accessible to tractors and large farm machinery. Portal-framed barns for livestock will usually have ventilated timber boarding on walls (known

If the structure is sound, a Dutch barn can be renovated, and even recovered with tin if required, to make a useful hay store or sheep barn.

as Yorkshire boarding), but ventilated steel sheets on the walls are an alternative. Modern livestock buildings will include a ventilated ridge line. Roofs are usually fibre cement or asbestos sheets. These are better than metal for livestock housing as they do not drip condensation in winter like metal roofing sheets do.

Dutch Barns

In recent years, many of these curved metal-roofed structures have been pulled down or replaced as they reach the end of their useful life. At the same time, Dutch barns are increasingly being appreciated for their aesthetics by non-farming folk taking on old farms. It is possible to renovate Dutch barns to make them safe and useful again, and they are a valuable addition to any farm or smallholding. The roof usually spans between 4.5 and 7.5m (15–25ft), giving a useful space under cover. The height of the roof makes the barns useful for stacking hay or straw and parking tractors. These barns also make good livestock housing.

BUILDING SAFETY

Check that whatever buildings you use are safe, both for the sheep and the people looking after them. Remove broken gates and barriers that have sharp edges. Fix loose slates, doors and roofing sheets, or anything else which could be caught by the wind. Remember that farm building roofs, especially fibre cement and asbestos, can be very fragile and dangerous so take safety precautions.

You will be visiting the lambing shed often, sometimes in the dark. When you have not had enough sleep, a tidy yard and building makes a much safer place to work.

Farm Asbestos

Asbestos and fibre cement roofs can be extremely dangerous to work on as they are fragile. Work should only be undertaken by experienced, insured contractors using necessary safety measures – such as safety nets underneath and crawling boards.

The hazards to health of breathing in asbestos dust are well known. Farms in the UK often have old asbestos sheets or pipes lying around; asbestos was not completely banned in the UK as a building material until 1999, so farm buildings erected in the 1960s, 70s and 80s will often have asbestos roofs and sometimes internal dividing walls. These are best left undisturbed or cleared away by specialist contractors. Avoid breaking or cutting the sheets or doing anything that will cause them to release dust. Modern fibre-cement roof sheets look like asbestos but, if manufactured after 2000, they will not contain asbestos.

CHAPTER 10

NUTRITION

Concentrate feeds such as straight cereals and sheep nuts should be fed carefully to avoid causing acidosis of the rumen. Before feeding, weigh the feed and use a scoop to measure the ration before filling a bucket. Calculate how many grams of feed you need for the type of sheep being fed: for example, a ewe suckling two lambs will need more than a ewe that is feeding a single lamb.

Forage or fodder means food for ruminant animals, like grass and hay. For a healthy rumen, forage must make up the bulk of the sheep's diet.

TYPES OF FORAGE

The simplest and best forage for sheep is grazed grass. In winter, the nutritional quality of grass declines, so shepherds will offer pregnant ewes an alternative 'conserved' forage such as hay, silage or haylage.

Hay

Hay is dried and baled summer grass, with a moisture content of less than 18 per cent. The quality of hay is important, so take time to assess the smell, colour and texture of hay that you buy. Ideally it will be more leaf than stalk. Good hay smells sweet and not musty and still has a slight green colouring to it. Sheep do not like old 'stalky' grass made into hay in July/ August but prefer the younger, leafier type made earlier in the season (June or early July). If offered poor-quality hay, sheep will eat less of it and they will lose body condition. Hay can be difficult to make in the wetter western areas of the UK, as five consecutive dry days are needed to dry the crop.

Silage

For silage, standing grass is mowed and then wilted in the sun. Silage is baled at a higher moisture content than hay and preserved through anaerobic fermentation by excluding air with plastic wrapping. Naturally occurring micro-organisms produce natural acids that 'pickle' the grass. Once silage bales

In Britain, a few days of bright sunshine close to midsummer are required to dry grass sufficiently to preserve it as hay. Round bales are labour saving, as a large volume of forage can be moved with a tractor.

are opened and the silage is exposed to the air, it needs to be eaten within two days or it will heat up and go off. Big bales of silage are only suitable for larger groups of sheep, around forty-five head plus. Good, leafy silage is very palatable to sheep as it is a moist feed and generally higher in protein than hay.

Haylage

Haylage is made like silage but the grass is turned after mowing to reduce the moisture content. The crop is 'tedded out' with a tractor-mounted machine called a 'tedder' or 'hay bob' that uses moving forks to aerate the grass, so it dries out more quickly. It's usually baled after one to two days of drying in sunshine and wrapped in plastic to exclude oxygen; haylage will go mouldy if exposed to air.

WHICH FORAGE FEED?

Advantages of Silage and Haylage
- Can be stored outside as it is wrapped in plastic
- Usually more palatable (tasty) to sheep, encouraging higher intakes
- Quicker and easier to make than hay as the mown grass does not need to be as dry before it is baled
- Higher protein content than hay because it is usually harvested earlier in the season, and clover in the crop is not over-dried, causing the leaf to shatter

Advantages of Hay
- Can be harvested with a small conventional baler (suitable for small tractors) – no need for hiring contractors or larger tractors – and made into small bales of hay that are easy to handle. Small bales are ideal for feeding to smaller groups of sheep

- Possible to make or buy big bales of hay. The hay will not go off when the bale is opened if it is kept dry
- No plastic required
- Better for birds, as the crop is turned and dried, leaving seed on the field for the birds

A big bale of hay being fed to ewes to supplement winter-grazed grass.

It is possible to make small bales of haylage by wrapping conventional small hay bales in plastic. This process is heavy work, however, and requires a specialist wrapping machine. You may find it for sale in your area but it is more expensive than hay. On the plus side, it can be stored outside and is not dusty like hay.

Alternative Forage Plants

Straw Sheep will eat the higher-quality forms of oat, barley and pea straw but require grass or hay and concentrate feed in addition.

Trees and shrubs Given the opportunity, sheep will graze on the leaves of trees and hedgerow plants, providing them with a mineral-rich added source of fodder. Among the leaves sheep like to eat are sycamore, willow, ash, ivy, lime and young beech.

Herbs and clover Herb seeds can be added to grass seed mixtures if re-seeding pasture. Useful varieties include plantain or 'ribgrass', sheep's parsley and chicory. Herbs and clover improve the mineral and protein content of pasture, helping to keep sheep healthy and growing well. Chicory is deep rooting and high in tannins, which can inhibit the development of internal parasites.

A mixed crop of turnips, forage rape and kale can be sown in mid- to late summer and will provide a valuable source of carbohydrate and protein to sheep in winter.

Roots and brassicas Sheep farmers often grow swedes, turnips or rape and kale for winter grazing, often sowing them as a mixture in summer and allowing the sheep to graze the crop from November onwards once the quality and quantity of grass has declined. Swedes are higher in carbohydrate and sugars than turnips but take longer to grow – a crop of turnips can be grown in just twelve weeks from mid- to late summer. Hay or a grass section of a field is provided to sheep grazing fodder crops to balance the sheep's diet.

Sheep grazing a fodder crop of turnips and kale on a cold winter's morning.

CHEWING THE CUD

It is a satisfying sight watching your sheep lying down and relaxing while merrily chewing the cud, or 'ruminating'. For healthy digestion of forage, the sheep will spend several hours a day regurgitating and chewing the forage eaten earlier.

If a sheep is not 'cudding' while all her flock mates are, this can indicate she has a health problem. If she is obviously under the weather, taking the ewe's temperature can be a useful first step. Other signs of ill health may be lethargy, not being able to stand, not eating and drinking, limping and diarrhoea.

Ensure your housed ewes always have access to hay or haylage/silage and an ample supply of fresh water. When housed and eating dry forage, pregnant ewes require more fresh water than when living off grazed grass.

THE RUMEN

The rumen is the first stomach (of four) and a miraculous synergy of mammal and billions of micro-organisms working together to digest high-fibre rough forage.

Forage should be available ad lib. A steady supply of forage is vital to keep the sheep's rumen and all the micro-organisms ticking over happily.

UNDERSTANDING CONCENTRATES

Concentrate is often referred to by farmers as 'cake' – not a Victoria sponge, but a blended food that contains a high amount of cereals. Cake usually comes in pelleted form, often called 'sheep nuts' and is available from all agricultural merchants.

Animal feeds are usually made of by-products from manufacturing. For example 'oat feed' or 'wheat feed' on the label means the sheep feed includes what is left over after oats and wheat have been milled and the best portion used for human

consumption. The label may also include the word 'expeller', for example in 'soya expeller', which indicates a by-product instead of whole soya bean. Try to find feeds with the greatest percentage of whole ingredients.

Concentrated sheep feeds are nutritionally balanced with a ration of protein (around 18 per cent), carbohydrate, fibre and minerals. The amount of useful carbohydrate is expressed as ME (measured in megajoules per kilogramme), which means metabolisable energy, or how much of the carbohydrate in the feed is useful to the sheep as a source of energy. ME is also relevant in the analysis of forage. Agricultural merchants usually offer farmers a service for analysing hay and silage to determine the nutritional content. The farmer will then balance the diet with an appropriate purchased feed. It is not recommended to feed pregnant ewes concentrate feed which has an ME below 12MJ/kg of dry matter.

Protein is expressed as crude protein on the label, and the quality of protein is important. The protein described as DUP (digestible undegradable protein) is the type that will benefit the ewe's colostrum production and

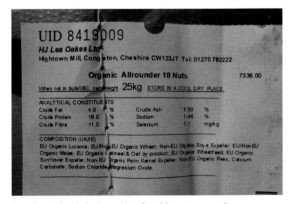

Look at the label on the feed bag to see the ingredients. A good-quality feed will contain some wholegrain ingredients, such as barley, oats, wheat or whole beans and peas.

lamb growth. Soya is the usual choice by feed companies to add quality protein to concentrate feed but it is imported and the use of it is questionable in the context of climate change and sustainability.

Alternatives to Soya

Various alternative protein crops can be homegrown or bought in for your flock from UK producers.

Clover silage Both white and red clovers offer a useful addition to any grass-seed mixtures sown on your land by boosting the protein content of grazing or forage. A newly sown sward with a high clover content can provide around 19 per cent crude protein. Red clover is highly productive and can be grazed or baled. Do not feed a pure red clover sward or bale to sheep six weeks before or after tupping time, as it can disrupt the ewes' oestrogen levels and affect their pregnancy.

Peas and beans Many forage varieties of peas and beans are available for growing in the UK as livestock feed. Organic farmers often grow a mixture of cereals, such as oats or barley, with peas or beans. The crop is harvested by a combine harvester and stored for winter. The mixture of cereal for energy and pulses for protein can provide a complete home-mixed ration for pregnant ewes or fattening lambs.

Alfalfa, lucerne and lupins These are sometimes included in organic concentrate feeds to add protein as an alternative to soya. These crops are usually grown in Europe or the east of England, where it is warmer and drier.

Forage rape and kale When grown as a mix with turnips or swedes, these plants can provide sheep with a homegrown winter diet rich in carbohydrates and protein. Twin-bearing ewes will usually need supplementing with concentrate feed or straight cereals closer to lambing, but the forage crop will prevent them from losing weight during the middle stage of pregnancy. Root crops also provide a useful break from grass, disrupting pest and disease cycles, and are a good opportunity to re-seed pasture.

Risks of Concentrates

Copper Poisoning

Some breeds of sheep (for example Soay, Texel and other short-wool continental breeds) are susceptible to copper poisoning if copper builds up in a sheep's liver over time. It's usually a problem in sheep or lambs being fed high rates of concentrate. Keep copper content in mind when checking the label of concentrate feed or selecting a source of mineral, such as a bolus or mineral lick. Do not feed 'nuts' (concentrate feed) to sheep that was intended for another species (like horses or cattle), as it may be higher in trace amounts of copper.

Copper toxicity is the most common form of poisoning in small domestic ruminants. Early signs include weakness and aimless wandering, progressing to jaundice (apparent in the whites of the eyes), rapid, shallow breathing and death.

Acidosis

Sheep love to eat concentrate feed and will eat it until they are ill. If sheep are fed too much concentrate containing cereals, especially wheat, they can be at risk of acidosis of the rumen. In extreme cases, this can cause death of the animal. To avoid this, the feed must be rationed – feed no more than 300g (10oz) per head at any given time. Ensure that dominant ewes don't eat more than their fair share, splitting them off into a different group if necessary. Make sure the sheep always have access to clean water and forage (grass and or hay, haylage or silage).

STRAIGHTS

Feeds are called straights when one unprocessed ingredient is fed to livestock. This is usually whole cereal grains such as wheat, oats or barley. It is possible to feed to sheep whole cereals such as oats or barley without any processing. Rolling the feed cracks the grains open and increases the digestibility. Rolled oats and barley are available in small bags from agricultural merchants.

If feeding straights to sheep, the diet will need balancing with protein and minerals. Protein can be provided with good-quality conserved forage or grazing; the more grass leaf and clover in the forage, the higher the protein content will be. The protein content of grass tends to rise in the early autumn and spring while the fibre and carbohydrate content may fall. This can be a reason for sheep scouring (diarrhoea) at these times of year. If concerned, ask a vet to analyse the dung to rule out internal parasites. In some feed blends, beans and peas are mixed in to provide protein. Another protein source can be a crop of kale and rape.

It is important that twin-bearing ewes in the last six weeks of pregnancy and six weeks after lambing have a well-balanced diet with good-quality forage plus adequate energy and protein in their diet.

FEED STORAGE

Good housekeeping around barns and yards is essential. Avoid leaving any spilled food on the floor, and make sure rats, mice and birds cannot access your feed stores. Store sheep food in rat-proof containers, and keep an eye out for signs of rat activity around the yard – if rats find a reliable food source, they will reproduce and the population increase rapidly. Rats will also burrow into

Home-produced grain, consisting of oats and peas grown as a mixture in the field, harvested by a combine harvester and processed through a roller mill to crush before feeding.

straw stacks if there is cereal grain inside the bales.

It is essential that hay is kept dry and out of reach of rain. Store the bottom layer of hay bales off the floor by stacking them on wooden pallets. Silage and haylage bales should not have any holes in the plastic. Discard any mouldy silage or musty hay to the muck or compost heap.

MINERAL SUPPLEMENTS

Buckets and Feed Blocks

Mineral lick buckets usually contain a mix of molasses - for energy - and minerals. Organic versions are available.

High-protein feed blocks can be bought from merchants, but these are not available as certified organic versions as they usually contain fish oils, genetically modified soya and maize. Feed blocks are useful when sheep farmers want to supplement the ewes while they are still out at grass. Access to a feed block and hay in the field can give the sheep a balanced diet pre-lambing.

A downside of mineral buckets and feed blocks is that it can become very muddy as the sheep gather around them. Moving the blocks or buckets every day can help. Another problem can be that shy or timid individual sheep may not use the block or bucket lick.

Mineral Boluses

An alternative way to supplement sheep with trace minerals is to use a mineral bolus. This is like a large tablet that the sheep swallows. The bolus sits in the rumen and slowly dissolves over a few months. Many soils in Britain are deficient in selenium and cobalt, trace elements that are important for growth and the immune system, and boluses are a useful way to ensure each sheep receives a

supplement. Be patient when learning how to administer them, as it does take some practice.

SUSTAINABILITY CONCERNS

Over the past three to four decades, conventionally produced animal feeds in the UK have generally contained imported ingredients, such as maize and soya - the feed label will state this. The sustainability of using these imported grains has long been

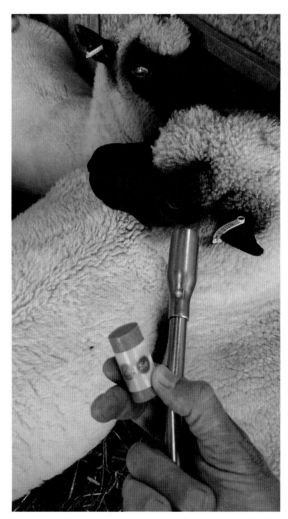

A mineral bolus is given to the ewes with a special applicator that carefully pushes it down the throat.

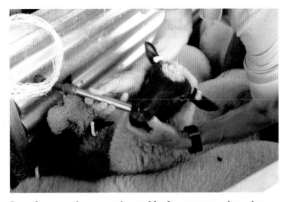

Lambs are given a mineral bolus at weaning time (four to five months of age). Start by putting the tip of the bolus 'gun' into the side of the mouth (where there are no teeth). Care needs to be taken not to hurt the sides of the mouth or throat when administering the bolus. Follow the instructions that come with the product.

questioned by proponents of organic farming and environmentalists, mainly because they are often grown on land recently cleared of virgin forests. It is simply not an efficient use of resources to grow arable crops on one side of the planet and then transport them to Europe to feed to animals that extract only a fraction of the nutritive value. It is possible to rear livestock in the UK without the use of imported ingredients from far-off continents.

Genetically Modified Feeds

Another issue with imported soya and maize is that it is often produced from genetically modified seeds. Check the feed labels on the bag to find out. GM maize and soya are bred to be resistant to glyphosate weedkiller so the resultant feed will contain traces of glyphosate as well as modified genes. These crops have long raised safety and ethical concerns with environmentalists and organic farmers. The only way to avoid GM feeds is to buy certified organic feed or local grain where you can be sure of its provenance, or grow your own. On our farm in Wales we grow a mix of organic oats and peas, which provide a good source of energy and protein for the pregnant ewes.

A crop of turnips and kale sown in midsummer can grow rapidly and provide a good homegrown winter feed. Pregnant ewes should be offered hay and, if possible, a grass run-back area to lie on.

CHAPTER 11

FEEDING SHEEP IN WINTER

Feeding a homegrown mix of rolled oats and peas to pregnant ewes in December.

FEEDING EWES FOR HEALTHY PREGNANCY

If close attention is given to how ewes are fed through the different stages of pregnancy, it can make lambing much smoother.

Four main factors determine how much extra feed a pregnant ewe should be given to supplement grazing or forage:

- The ewe's body condition
- The size of ewe
- How many lambs she is carrying (litter size)
- The stage of pregnancy she is at

Monitor the ewes with these factors in mind throughout the gestation (pregnancy), which will last on average twenty-one weeks (five months).

Ideally, ewes should be body condition score (BCS) 2.5 or above before they are put with a ram. It is undesirable to feed thin ewes extra rations during tupping time, as this will only encourage their bodies to produce more lambs. Thinner ewes struggle to produce twin lambs of an adequate birth weight and will not produce enough milk to feed them. The aim should be to have ewes in good body condition before tupping time begins.

There's no one-size-fits-all diet for sheep, and a good shepherd will be constantly thinking about the condition the sheep are in, tweaking the feed rations and ensuring there is a constant supply of good-quality pasture available for the sheep going forward.

Speaking to other breeders with the same type of sheep can help you formulate a diet for your ewes. Keep in mind, however, that people who keep show sheep will have very different feeding regimes to those who farm sheep commercially or who keep them as pets.

Body Condition

Assess the ewes' body condition five to six weeks after the date when the rams were first introduced to them. This can coincide with taking the ram away from the breeding flock. Try to determine if the ewes have lost weight or not.

If the ewes have lost weight during the tupping period (to BCS 2 or under), look at the quality of their grazing and test to see if they're suffering from internal parasites. Consider starting to feed them once a day with concentrate or straights; up to 200g (7oz) per head per day should be ample for medium-size ewes in the first half or middle of pregnancy.

Adjust your feed rate according to the size of the ewe. If the ewes are short of grazing, provide a covered hay rack and start feeding hay or haylage. Request that a vet checks the ewes' dung for liver fluke and worms, as the presence of these parasites can lead to dramatic weight loss. Older, broken-mouthed ewes (with teeth missing) will not be able to maintain their body condition over winter with forage alone. Getting the health and body condition of ewes right before and after tupping is vital for a successful lambing time in the spring.

Overweight Ewes

Ewes that are above BCS 3 at tupping time can be allowed to lose weight during the first half of the gestation period (autumn/early winter for spring-lambing sheep). However, ensure that any weight loss is not due to illness from worms or liver fluke.

Size of the Ewe

The size of ewe will depend on the breed of sheep or type of cross-breed that you decide

FEEDING BY BREED AND LOCATION

The altitude, aspect and location of the farm are all factors in determining how the sheep are fed, as well as the breed of sheep kept. Hill breeds are exceedingly hardy, and, if kept on lowland farmland, may not need any supplementation. Conversely, larger breeds, such as Suffolk, Dorset or Shropshire, will need extra feed to supplement grazing if kept on higher, exposed ground. Learn as much as you can about feeding sheep through the seasons from local sheep farmers and see what breeds or cross-breeds are popular in the locality.

to keep. According to the National Sheep Association, there are over ninety different sheep breeds in the UK, with very different characteristics, feeding requirements and body weights. For example, at the extreme ends of the scale, a 'primitive' Soay sheep will have a mature female weight of around 24kg (53lb) and likely only carry one lamb, while a mature Suffolk ewe can be anything from 80kg to 160kg (176–352lb) and is more likely to have twins or triplets. Somewhere in between, a Lleyn ewe would be around 50–60kg (110–132lb), and 70 per cent of a Lleyn flock would usually carry twins or triplets.

Weather

The weather can play a big role in how the sheep are faring. Persistent wind and rain over the course of a month can cause weight loss, as the ewes will be using the energy from their feed to keep warm. In very cold periods with frost and snow, the sheep will require extra forage and maybe concentrate feed.

Feeding amounts need to be adjusted for the type of breed or cross-breed. Knowing the type of ewe's average mature weight can help you calculate concentrate feed rations.

Pregnancy Scanning

Scanning is a useful tool for the management of pregnant sheep. For a successful lambing, it is vital to feed the ewes appropriately according to their body condition and how many lambs they are carrying. Scanning the ewes allows you to separate them into groups of 'singles', 'doubles', 'triplets' and 'quads', and feed to their energy and protein requirements. The scanning person will usually put a paint mark on each ewe's fleece to indicate how many lambs she is carrying, for example blue dots for ewes with single lambs, red for twins and so on.

Sheep scanning is a specialised skill and carried out by a contractor who travels from farm to farm. Ask sheep farmers in your area for a contact. Cost will usually be charged per ewe plus a set-up/travelling fee for smaller flocks.

Scanning has to be carried out at the right time; if left too late, the foetuses can be too large and the operator cannot see how many

An example of a sheep ultrasound scan.

Sheep ultrasound scanning in progress. The contractor sits in a dark box with the ultrasound machine screen. The sheep walks up the ramp and stands next to the operator; the operator holds the scanning wand against the underside of the ewe's belly and then puts a paint mark on the side of the ewe to denote single lamb, twin lambs, triplets or if she is empty and not in lamb. With experienced sheep-scanning operators, the process only takes about thirty seconds per sheep.

a ewe is carrying. The timing is calculated by the number of days since the ram was first introduced to the ewes. The optimum time for scanning is eighty to ninety days after the ram was turned out.

Pregnancy scanning early in gestation allows the shepherd to feed the sheep according to their needs. This can save a lot of problems when it comes to lambing time. Properly fed sheep with singles will not have over-large lambs that are difficult to birth, and sheep with twins will not give birth to weak, underweight lambs and be short of colostrum. A reduction in lamb losses and more accurate use of feed supplies will also save the shepherd money.

Feeding Ewes in Late Pregnancy

Around 70 per cent of the growth of a lamb foetus is during the last six weeks of pregnancy (the third trimester). This rapid growth puts large demands upon the mother.

If a ewe's requirement for energy and protein is not met during this time, she will become thin and her lambs may be born underweight and weak without enough fat for them to metabolise in the first hours after birth. An under-nourished ewe may also be short of colostrum and milk for her lamb(s). If the ewes have been scanned, the shepherd can take account of the litter size as well as the size of the ewe when formulating a daily ration of concentrate feed.

Ewes Carrying Single Lambs

Ewes carrying a single lamb will usually produce a healthy, strong lamb and plenty of milk to feed it without additional concentrate feed. This does depend on having good-quality hay, silage or grazing and supplementary minerals. In flocks where the sheep are housed pre-lambing, the singles are often left out in the field and either brought in just before lambing starts or lambed outdoors in the field. Elderly ewes in poorer condition with missing teeth may need supplementing with concentrate.

Single-bearing ewes will perform well on grass if supply is adequate. If the ewes

Whether indoors or outdoors, ensure twin-bearing ewes receive adequate nutrition in the last eight weeks of pregnancy. Offer ad lib hay or silage and calculate concentrate feed amounts according to the average ewe's weight and how many lambs they are carrying. For better rumen function, split the ration into two feeds – early morning and late afternoon.

are housed, or grass is in short supply in the field, ad lib forage (hay or silage) and access to a mineral lick is usually sufficient. If you have a policy of trying to foster spare lambs onto ewes with singles, it is worth paying attention to the ewe's nutrition so you can ensure she will have enough milk available to feed two lambs. However, be very careful not to overfeed single-bearing ewes, as this can lead to large lambs and difficult births. If unsure about the quality of your hay and silage, ask an agricultural merchant about forage sampling. The results will tell you the levels of metabolisable energy (ME) and protein in the forage (*see* Chapter 10).

Feeding Ewes with Twins or Triplets

As a guide, a ewe weighing 70kg (154lb) carrying twin lambs will need between 1.7 and 1.9kg (3.7–4.2lb) of dry feed per day. An ample supply of grazed grass and forage will supply this in the first two trimesters of pregnancy. In the third trimester (the last six weeks before lambing), the rumen is squeezed by the growing lambs so the ewe cannot take as much forage on board. During this period it is important she has good-quality forage: in Britain, this is grass that is made into hay or silage in May or June (unless it is a late sown new ley, or two cuts are taken in the season). Silage or haylage with a high clover content will be much higher in protein and can allow for less concentrate feeding.

The table below gives a typical concentrate feeding regimen for the last eight weeks of pregnancy for twin-bearing ewes. This is on top of the ad lib forage provided for all sheep.

As a rule of thumb, the ewes should eat all the concentrate feed given within two to three minutes. If feed is left in the trough, it has either been contaminated with mould, damp or rat pee, or the sheep are being overfed.

WINTER DIETS FOR OTHER SHEEP

Grass is the food of choice for all types of sheep but, at certain times of the year, the

Amount of concentrate feed per head in last eight weeks of gestation

Weight of ewe	Eight and seven weeks before lambing (in a single feed)	Last six weeks of gestation (in two feeds, early morning and late afternoon)
50kg (110lb)	250g (9oz)	250g (9oz)
70kg (154lb)	300g (10.5oz)	350g (12oz)
90kg (198lb)	400g (14oz)	450g (16oz)

sheep's diet will need to be supplemented with forage or a concentrate feed containing higher energy and protein. Offer hay to the sheep in wintry conditions. If they do not need it, they will leave it uneaten. If the area of grass is limited, you may decide to house all your sheep during some of the winter months in order to give the ground a rest and let the grass recover.

Growing Lambs

From weaning to slaughter at six to twelve months, lambs should be grass-fed with mineral supplementation if required. In winter, forage and maybe some concentrate will be required for fattening or because of limited grass.

Ewe Lambs as Flock Replacements

With rotational grazing over winter, ewe lambs will not usually need extra feed unless there is snow or severe cold, when they will require supplementing with hay or silage. Monitor for internal parasites and treat if required. Their mineral intake can be supported by providing licks, but consider consulting a vet first to establish if this is necessary for grass-fed lambs. This can be done by blood testing a proportion of the flock.

Breeding Rams

Unless grass is in short supply or the rams are exposed to harsh winter conditions, they should prosper on pasture alone all year round. If rams are being prepared for traditional show and sale, then a specialised diet containing concentrate and forage will be required. This type of diet can cause health problems, however, and some farmers prefer to buy rams that have been exclusively grass reared, even though they will not look as impressive as rams bulked out with concentrate feed.

Pet Sheep

After weaning, lambs reared by bottle feeding can be fed a handful of concentrate daily and left to graze. Offer hay or silage in wintry conditions. Monitor bottle-reared sheep closely for weight loss, as they are more susceptible to internal parasites and disease than lambs naturally reared by the mother. When they reach a mature size, they will not require supplementary feeding unless the grass is in short supply.

Hill and Primitive Breeds

Consult other breeders or the relevant breed society for advice on supplementary feeding of small primitive breeds like the Manx Loghtan, Soay, Hebridean and Balwen.

A group of strong and healthy ewe lambs in spring after being overwintered on pasture without supplementary feed.

Unless they are going to be prepared for a show or sale, rams will usually thrive over winter on reasonable-quality pasture without needing supplementary feed.

These breeds are exceptionally hardy and can survive on poor-quality and sparse grazing. Being very good at escaping, some of these breeds can be difficult to contain, so consider feeding a little concentrate to discourage them from wandering. They are more likely to have a single lamb. Primitive breeds that carry twin lambs and are kept in their natural habitat of rough or sparse grazing will usually require some supplementary feed.

Moving ewes to fresh pasture in November. If possible, rotate the ewes around the grazing, moving them every week until they have eaten all of the residual autumn grass. In a mild winter, the grass will still grow a little if the pasture is rested from grazing.

Three hoggets that were bottle reared from birth and fed daily with a small amount of concentrate.

CHAPTER 12

LAMBING TIME

A fresh lamb born outside in April.

In a breeding flock of sheep, the whole calendar year is shaped by the date the shepherd chooses to start lambing. The lambing period for most British sheep breeds is between January at the earliest and May at the latest. The Dorset breed is an exception to this, as the ewes can lamb at any time of year and are able to conceive in the spring and give birth in the autumn. For the majority of breeds, however, the start of a pregnancy is seasonal, with the ewes only beginning to ovulate and come into season as the days shorten in late summer.

GESTATION LENGTH

The average gestation length for sheep is 147 days (about five months). This can vary by breed, topography and climate by a few days either side. If you know what date the ram served the ewe, you can be reasonably accurate in predicting when she will lamb: clear your diary for that week or book some annual leave. For April lambing, for example, the ram is turned out with the ewes on 5 November, so the first lambs will appear on 1 April .

EARLY LAMBING

Some farmers choose to lamb early in the year (December to mid-February) because they want to have lambs ready to sell at Easter to take advantage of higher prices: traditionally, new-season lambs sold for meat at Easter will achieve the best price per kilo. Even later in the season, those extra weeks of growing time will be an advantage, with the lambs returning a higher price per kilo than ones sold for meat later in the season.

If you plan to show your sheep at agricultural and breed society summer shows, earlier lambing gives the opportunity to present a larger, more-impressive looking animal with more wool on its back than later-born lambs.

Finally, lambs born earlier in the year are generally more resistant to the problems posed by internal and external parasites in the summer months and are less likely to develop clinical disease than lambs born later (April or May).

Disadvantages of Early Lambing

Early lambing has disadvantages, using resources of labour, feed, medicines and housing more intensively:

- Early lambing beginning in December relies on artificial hormone treatment to bring the ewes into season.
- An early-lambing flock requires adequate housing space per ewe. If suitable barns are not available, this can be a significant cost at the outset.
- Inputs of labour feeding, spreading bedding and mucking out are greater than for an April outdoor lambing flock.
- The cost of feed is higher. As the grass lacks nutritional quality in the winter, heavily pregnant and lactating ewes with twin lambs will need supplementary concentrate feed to meet their energy and protein requirements.
- Good hygiene achieved with adequate quantities of clean straw is necessary for earlier indoor lambing to prevent illness in newborn lambs such as 'watery mouth' (*E. coli* infection) and mastitis in ewes.

Sheep lambed indoors require an ample supply of clean, dry straw for bedding. Top the straw up daily to keep the ewes and lambs clean. Dirty and wet bedding can lead to bacterial infections in ewes and lambs.

INDOOR OR OUTDOOR LAMBING?

There are many reasons for choosing one system over another and the decision comes down to what best suits your breed of sheep, the objectives of your sheep enterprise, the buildings available, the geographical location of your farm or smallholding and your willingness to brave the elements and wander the fields in the middle of the night, checking on your ewes by torchlight.

Indoor Lambing

Advantages

While the sheep are housed, often six to eight weeks before lambing, the grassland will have a break from grazing. This means there will be more grass available for the ewes with lambs at foot when they return to the fields in springtime.

Ewes scanned with twins or triplets can be housed in a separate pen from ewes with single lambs and fed according to how many lambs they are carrying. Dividing the ewes into groups post scanning helps to prevent weak and underweight twin lambs being born. At the same time, ewes with single lambs will not be overfed, preventing the unborn lambs from becoming too large. Inside a barn, the ewes are sheltered and protected from wind, rain and snow during the time they are heavily pregnant.

The ewes can also be checked more quickly indoors at nighttime without long walks in the dark around the fields.

Indoors a ewe needing assistance during the birth is easier to catch. Similarly once the ewe has lambed, she can easily be put inside a smaller pen with her lambs. The shepherd can then check the lambs are receiving adequate milk before they are released into the field.

Other husbandry tasks, such as drenching the ewe for worms, trimming her feet, tagging the lambs, docking their tails and applying elastrator rings to the male lambs can all be done inside the barn before turn-out. These tasks are easier under cover from the wind and rain and without the need to chase and catch the ewe or her lambs in a field.

Disadvantages

To successfully lamb sheep indoors, good housekeeping in the lambing shed is

It is easier to check sheep and deal with problems promptly if they are housed when close to lambing time, rather than you having to walk the fields in the dark.

essential. This means keeping the bedded areas and water buckets clean to prevent the spread of infection. The sheep will need their hay racks filling up regularly, clean straw spread for lying on and water renewing to keep it fresh. Small lambing pens must be cleaned out between each ewe and her lambs.

In overcrowded, poorly ventilated, dirty sheds or barns, lambs are prone to die from bacterial infections such as *E. coli* (commonly called watery mouth), navel infections and joint ill (infections of the joints). Foot infections can also spread between ewes more easily if the ewes are not footbathed and affected individuals separated and treated before the housing period begins.

Ewes can become unfit and overweight while indoors, making lambing problems and prolapses more likely.

If you cannot use an existing building, the costs of providing a barn for your sheep can be prohibitive for a small flock. Expenditure will also be required to fit the lambing space out with sheep hurdles, feeders, hay baskets and water troughs, plus the need to purchase straw for bedding.

Outdoor Lambing

Advantages

The lambs are born into a clean natural environment. The ewe can wander the field during early labour and choose a spot where she is comfortable to give birth. Some breeds of sheep, particularly hill breeds, will be much happier in themselves to remain outdoors all winter. As long as they are checked up on regularly and offered good quality hay in bad weather, they will be happy.

Outdoor-lambing ewes should be generally fitter and give birth easier than ewes that have been housed for six to eight weeks. The risk of vaginal prolapse for twin-bearing ewes is much less with outdoor-lambing flocks, as long as they are not overfed with concentrate feed.

With good grass cover on the field, the risk of infection and joint ill via navel infections in the lamb is much reduced.

Savings can be made with outdoor lambing as there is less need for straw and lambing hurdles.

Disadvantages

Of course, March and April weather can bring the possibility of wintry conditions with sleet, hail and driving rain, particularly in mountainous areas. Rough weather can put newborn lambs in peril from hypothermia and death by exposure.

Outdoors a ewe also has a greater risk of losing a lamb to predators such as foxes, crows and badgers before the shepherd has a chance to help them. Lambs rejected by the mother at birth will certainly be lost unless found by the shepherd soon after birth. Sheep that need help during the birth are often difficult to catch in the field, and working alone in the field night after night can be exhausting for the shepherd, especially in bad weather.

Outdoor-lambing flocks tend to lamb later in the season to avoid the worst ravages of wintry storms. By midsummer, lambs born later (mid-April onwards) are smaller than their late winter- or early spring-born

A Llanwenog ewe cleaning her newborn lamb after giving birth in the field in April. In good weather, ewes are happier lambing outdoors in their natural environment.

counterparts. Pests and diseases associated with warm summer weather can affect the smaller lambs more than larger, older ones. Furthermore, the later-born lambs may not be ready to sell by the following winter and the hoggets will then be competing with the breeding ewes for dwindling grass supplies.

LAMBING ESSENTIALS

- Iodine spray – for spraying the lambs' navels as soon after birth as possible to prevent infection
- Bar of soap and nail brush for washing your hands thoroughly before assisting a ewe
- A clean bucket and disinfectant to disinfect your hands before examining a ewe
- Calving gloves (optional); these are long, disposable plastic gloves for use when assisting a ewe
- Lambing aid for easing the lamb's head forwards in difficult births; this is a short piece of plastic-coated washing line or electrical flex. A loop is formed by pushing the two ends through a 2.5cm (1in) piece of alkathene plastic pipe (or similar plastic tubing)
- Lambing lubricant, for putting on your hand before examination of the ewe. Sometimes useful for delivering a large lamb where the head is stuck
- A bottle of antibiotic and anti-inflammatory medication (obtained from a vet) to give to ewes where the shepherd has had to assist by putting their hand inside the sheep to deliver the lambs. If you have access to a vet nearby, the medicine could be obtained as and when required without purchasing a whole bottle
- Anti-inflammatory medication, given to ewes on vet's advice as a painkiller after a difficult or prolonged lambing

Useful kit at lambing time (left to right from top): infra-red heat lamp, iodine, disinfectant, calving examination gloves, stomach tube kit, feeding bottle, colostrum replacement powder, clean bucket for warm water and soap.

Newly born triplet lambs looking for their first drink of essential colostrum. A shepherd should spend time with the ewe and lambs to ensure they all suckle and receive a belly full of colostrum.

- Colostrum, the thick, custard-like milk the ewe first produces for her lambs. Colostrum is energy-dense and full of antibodies that are the building blocks of the lamb's immune system. Natural colostrum can be milked from a ewe and frozen in an ice cube tray for later use, or powdered colostrum can be purchased in sachets for lambs whose mothers do not have any
- A secure pen made with hurdles in a corner of the barn or field where a ewe can be caught and the birth assisted if necessary
- The vet's phone number. Occasionally, when you have tried to assist the ewe by yourself unsuccessfully, calling the vet is the right thing to do. Country vets are experienced at dealing with complications in lambing and may make the difference between having a live lamb and ewe or not

PREPARING FOR BIRTH

When lambing time arrives, you should aim to check on the ewes at least every four hours.

This will give the best chance of saving lambs should complications occur. The majority of ewes will give birth without assistance but it is still useful to be there so that you can ensure the lambs are receiving colostrum and are being cleaned and cared for by the ewe.

In the last four weeks of pregnancy, you should notice development of the ewe's udder, which slowly fills and swells with colostrum ready for the newborn lambs. If no udder development is observed, this may mean that the ewe is not in fact in the last month of pregnancy, is under-nourished so she does not have the energy reserves to produce colostrum, or is not pregnant at all. If the ewes were scanned earlier in the gestation, this will help to determine the stage of pregnancy and when you can expect to see the production of colostrum in the udder.

Early Signs of Labour

It is not always easy to spot when a ewe is beginning the birth process but looking for these signs may help identify a ewe that is in labour:

- Restless walking around
- Getting up and lying down repeatedly

- Stops chewing the cud and eating and drinking
- Twitching, turning to look back at her flanks
- Repeated small bleats and sniffing the ground - often the ewe can smell her lambs from the amniotic fluid that begins to leak out so she will walk around calling and looking for them before she has given birth

Later Stages of Labour

If the birth is progressing normally, the ewe will continue with her distracted behaviour and then push out the water bag, either while standing up or lying on her side.

The water bag is a fluid-filled sack that the ewe expels to open the birth canal ready for the lamb. Once outside her body, the bag looks like a small balloon hanging down or, if the liquid has spilled out, like a string of white membrane. Ewes with twins or triplets will have two or three water bags.

Once the water bag has been delivered, the ewe should progress with labour and give birth within about forty minutes. The length of time can vary but if no part of the lamb is seen at the vaginal opening after thirty minutes this often indicates a problem requiring the shepherd's assistance. For young ewes lambing for the first time, the birthing will take a bit longer than a mature ewe that has lambed before.

In a normal delivery, the ewe will spend the last part of the labour 'straining', her sides moving in as she experiences strong contractions. Sometimes she will lie on her side while experiencing the contractions and appear to 'push' with her head up or prostrate on the floor. At other times she may have the contractions standing up, arching her back when pushing. At this stage of the birth you should expect to see some part of the lamb protruding or be able to feel it just inside the opening.

THE BIRTH

A good birth usually happens when the first lamb has its front toes just inside the opening with its nose just behind on top of the front feet. Usually the ewe should be able to push the lamb out with no assistance. A good shepherd will wait and watch quietly, intervening only when it is apparent the ewe is ceasing to make progress with the delivery on her own.

Sometimes a larger lamb will need a pull from the shepherd. Calmly approach the ewe while she is lying down and pull the feet of the lamb alternately. By pulling one foot at a time, you can 'wriggle' the shoulders through the ewe's pelvis, the narrowest point. Once the shoulders are through, the rest of the lamb will follow.

When to Help with Delivery?

This is a difficult question to answer, as experience is the best teacher. In any birth situation there are many variables at play, including the size of ewe and breed, the size and breed of the ram, the feeding of the ewe and her age and how many times she has given birth before. Here are some signs to look out for when deciding whether to intervene.

The restless, pacing behaviour of a ewe in pre-labour can sometimes go on for a few hours. Once the ewe gets to the stage of pushing and is obviously having contractions, make a mental note of the time. If the water bag has not been delivered within thirty minutes, this can indicate a 'malpresentation', such as the lamb coming bottom first (breech position). A ewe that is trying to deliver a lamb in a breech position does not usually progress to the stage of obviously pushing. If you notice that a ewe in labour seems to be stalling and not progressing, catch her in a small pen and examine the presentation of the lamb. If in doubt, or the neck of the uterus is not open, consult a veterinary surgeon.

If the ewe has passed the water bag out but a part of the lamb does not 'present' at the opening within thirty to forty minutes, this usually indicates a problem that requires investigation.

A ewe lambing for the first time will usually take ten to fifteen minutes longer at each stage of labour and can sometimes struggle to push out a big lamb without assistance.

Ewes having larger single lambs and presenting them normally (front feet and nose first) are more likely to need the shepherd to assist by pulling the lamb out by the front legs.

When faced with a ewe whose delivery has stalled, it is important not to panic: be methodical and find out as much about the position of the lamb as you can.

Examining a Ewe During Lambing
Catch the ewe and put her into a small pen. Ask your assistant to hold the ewe's head while she is either standing up or lying on her side. If she's lying down, keep her head up off the floor and a foot under her shoulder if you have a helper. Never hold the sheep by the legs or have her lying flat on her back, as she will struggle.

Sequence of pictures showing a ewe delivering a lamb naturally without assistance. Once the lamb is born, the shepherd can ensure the membrane sack is removed from over the lamb's nose or mouth so that it can begin to breathe.

The ewe's contractions are pushing the lamb out. The lamb's head and front feet are inside the membrane. There is no need to break the membrane until the lamb is fully delivered.

A ewe in the last stage of labour. She lies on her side as the contractions become powerful, and pushes the first twin lamb out.

Here the lamb is half out and it needs to be delivered quickly at this stage. The shepherd may need to assist by helping a larger lamb out.

Check to see if the membrane sack needs removing from the lamb's nose or mouth so that it can begin to breathe.

Whether using a glove or not, make sure to clean your hands thoroughly and cut your fingernails short.

Dip your hand in a bucket of warm, clean soapy water that has a splash of disinfectant in it. Hygiene is extremely important.

When examining a ewe, make your hand as long and narrow as possible, folding in your thumb towards your palm. Put lubricant on your hand and keep your fingernails away from the wall of the uterus in case they damage it.

With your fingers, try if you can feel any part of the lamb. If you can feel two front feet and the lamb's nose a bit further back, the ewe might allow you to pull the lamb, or she might need a bit more time to push herself.

Pulling a Lamb

Pull one foot at a time until the lamb's shoulders 'pop' through the narrowest part of the ewe's pelvis. When pulling a lamb by the front feet, always ensure the nose and head of the lamb is following and not stuck back.

Pull each time the ewe has a contraction. Once the lamb's head and neck are completely out, the lamb must be fully pulled out. When a lamb is half-way out, it can be starved of oxygen as the umbilical cord becomes pinched as the lamb's body passes through the mother's pelvis. If the lamb is large and you are not sure it can be delivered without damaging the ewe or lamb, consult a vet.

COMMON MALPRESENTATIONS AND HOW TO CORRECT THEM

One Front Leg Pulled Back

This is a fairly common presentation. The leg stuck back makes the lamb wider at the shoulders and this prevents it from sliding through the birth canal. If the lamb's leg and nose are not too far out of the uterus, there can be room to put your hand in, find the other leg and gently hook it forwards with your finger; you can then pull the lamb fully into the birth canal by its feet.

Head Tilted Back

One or both legs may be engaged in the birth canal but the head can sometimes still be back inside the uterus. To correct this, the lambing aid can be useful. Slide it in with your hand, aiming to get the loop of wire behind the lamb's ears. This manoeuvre can take some time but, once the wire is behind the lamb's head, your free hand can pull the feet of the lamb towards you while the other hand – outside the ewe – can ease the head along behind by pulling on the wire of the lambing aid. Note: with twins or triplets, it is possible to be pulling on the head and feet of separate lambs by mistake. Make sure the feet you are pulling on belong to the head and neck you can feel.

Back Feet First

When examining a ewe, try to ascertain if the feet you can feel are the front or the back feet of the lamb. The front feet of a lamb appear with the sole of the lamb's tiny hooves pointing down towards the floor (if the ewe was standing up) and the first leg joint will hinge to bend the foot towards the floor. Back feet are the other way round, with the sole of the hoof facing upwards and the first joint hinging so the foot moves upwards.

If you can feel both of the lamb's back feet, the lamb can be pulled out. Work slowly with the ewe's contractions to ease the lamb towards you. Once the lamb is out up to its middle you will need to pull it out firmly and swiftly. When the lamb is half-way out the umbilical cord will be pinched, as the belly of the lamb is in the birth canal. A prolonged wait at this point can starve the lamb of oxygen from its mother, and larger lambs can be lost during birth because of this.

Lambs delivered backwards are usually slower to start breathing and have more fluid in the airways. To help revive the lamb and get it breathing, pinch the middle part of its nose. Clear the lamb's mouth with your finger and push a piece of straw into its nostril to induce a sneeze reflex. If it is slow to start breathing, pick the lamb up by the back legs and swing it back and forth in a wide arc two to three times. The centrifugal force will help clear liquid from the lungs. Be careful not to bash it against any obstacles and ensure you have a firm grip. Some vigorous cajoling will usually spark a lamb into life, and you will not hurt it by doing this. The mother licking the lamb will also stimulate it to breathe.

Tail First and Back Legs Tucked Under

This is one of the most difficult lambing problems to correct for the novice. You will know if you have this presentation on your hands if you can feel the lamb's tail just inside the opening, or part of the lamb's spine across the opening to the uterus. Lambs will die in this position without help because the ewe's contractions will squeeze the lamb against her pelvis repeatedly.

If possible, have an assistant to hold the sheep's head, and position the ewe to lie on her side. The aim is to try to push the lamb back away from you to make room to bring the back legs round. Place your hand on the lamb's bottom and push it away from you further inside the sheep. This is not easy as you are working against the sheep's contractions.

If the lamb is not too large, pushing it back inside the ewe should make some room. Slide your hand down one of the lamb's back legs until you feel the large hock joint half-way down. With your finger around the hock try and 'tease' the leg towards you. Push the lamb's bottom away from you again and repeat. If this is altering the position of the lamb, continue by sliding your hand down inside, as if going towards the ewe's udder, until you feel the lamb's fetlock joint (the last joint closest to the foot). If possible, cup the foot of the lamb in your palm and bring it towards you. Do not pull the lamb's leg towards you from the hock joint, as this may make the lamb's foot rupture the wall of the uterus.

If successful in getting one of the back legs around, repeat the process with the other leg and you will be able to deliver the lamb as a back-feet-first delivery.

ASSISTING THE EWE INTERNALLY

- Keep a cool head and work methodically.
- Pay attention to hand hygiene, dipping your hand in the soapy disinfectant bucket if you have to pull it out and go back inside the ewe.
- Be careful not to let your fingers or the feet of the lamb damage the wall of the uterus.
- If you have had to put a hand inside the ewe to deliver lambs, aftercare of the ewe should include an anti-inflammatory painkiller and antibiotic injections. Follow the vet's instructions with regard to administration.

The lamb's head is out but the front legs are back inside the ewe. The lamb cannot be delivered in this position.

Head out but no Feet

In this situation, the lamb's head is hanging out of the ewe from the vulva but the front legs are stuck back, which can look rather dramatic when first discovered. The ewe's contractions have pushed the head down the birth canal and out of the vagina, so it looks like a swollen ball with eyes and a nose. I once had to chase a ewe around a field to catch her and help deliver the lamb, whose head was bobbing about beneath her tail like a football. I was convinced the lamb would be dead but was pleasantly surprised to find it was still alive. Once delivered and given mother's colostrum from a bottle, it thrived.

To deliver the lamb, you will have to push the head back inside the ewe and find the lamb's front legs. Using your bucket of warm, soapy water, wash the lamb's head and pick any bits of straw or dirt off. Put plenty of lubricant behind the lamb's ears and round the opening of the vulva. With your hand over the lamb's face and the nose in your palm, push it back inside the ewe as firmly as you can.

It will help to have the ewe lying on her side. It can also help to prop the ewe's back end up in the air with a lump of straw or hay and roll the ewe onto her back while you push the lamb's head back inside the ewe. Be prepared to apply firm pressure to pop the lamb's head back in, as you will be pushing against the ewe's contractions, making it difficult.

Once the lamb's head is back inside the ewe, locate the lamb's front feet and bring them around to the neck of the uterus. Pull the feet towards you one at a time.

If needed, use the lambing aid to ensure the head follows the feet and shoulder of the lamb during the delivery attempt. It may help to put the wire of the lambing aid behind the lamb's ears before pushing the head back inside the ewe, although this is not always possible.

Once delivered, the lamb will need extra care until the head and face swelling has

In this position, the lamb is still receiving oxygen via the umbilical cord and the placenta. To deliver the lamb, you will need to push the head back inside the ewe and try to find the front feet.

subsided. Help it to suckle from the mother or 'milk' some of the ewe's colostrum into a jug and feed the lamb with a bottle. Do this as soon as possible after birth.

Seeking Veterinary Help

If you feel that you are not going to be able to deliver the lambs, telephone the vet. Use it as an opportunity to ask questions and improve your experience so that next time you will be more prepared and confident to deal with it yourself. It is always worth having a small trailer with straw bedding ready for such a situation. Taking the sheep to the vet's practice will save the cost of a call out and usually it means the ewe will be looked at sooner.

CHAPTER 13

CARING FOR THE EWE AND HER LAMBS

Once the lamb is born, the shepherd's real work begins. It is important to ensure the ewe is bonding with both her lambs and is licking them clean. The action of the ewe's tongue cleans and dries the lamb, helping it to warm up.

Give first-time mothers time and space to bond with their lambs. Too much interference can cause them to reject one or both lambs, especially in hill breeds of sheep. A healthy lamb, born without assistance and carrying a good layer of subcutaneous fat, may look vulnerable but is very hardy. With a good feed of colostrum and an attentive mother, the lamb is surprisingly tough and able to withstand the cold and the elements. As soon as the shepherd is confident the lambs are being well fed and looked after by the ewe, she

A ewe cleaning her lamb and forming a strong maternal bond towards it.

and her lambs can be turned out into the field, although in wet and windy weather, delaying turnout by a day or two is a good idea.

COLD LAMBS

It is important to be vigilant and not let lambs become hungry and hypothermic after birth. This can happen quickly in cold conditions if the ewe is ignoring a lamb and not licking it clean, particularly to small and skinny lambs. Sometimes a lamb is hungry because the ewe will not let it suckle, and these are also at risk of hypothermia if left.

When a lamb is cold it cannot suckle. If unsure, use a thermometer to take the

The ewe licking the lamb stimulates the lamb to get to its feet and suckle from the mother's udder.

lamb's temperature, but you can often tell if a lamb is too cold by putting your finger in its mouth – if it feels lukewarm or colder than your own body temperature, the lamb will need help to warm up. A lamb's normal rectal temperature will be between 39 and 40°C (102–104°F). Shivering is the lamb's way of using its energy to warm the muscles up.

Occasionally it is necessary to intervene and rub down a lively but shivering lamb with a towel or straw. If this is not sufficient to get it suckling then it will need to be placed under a infra-red heat lamp until it is warm and dry. Give the lamb a feed of colostrum by bottle once it has warmed up, can stand and can suck with its mouth. Weak lambs that can swallow but not suck may need the first feed via a stomach tube.

Small, skinny lambs who are lively enough to stay with the ewe can benefit from having a little jumper fitted. The sleeve of an old fleece or pullover is ideal for this. Leave it on for twenty-four to forty-eight hours until the lamb is getting a belly full of milk regularly.

Signs of Hypothermia in Lambs

A lamb that is comatose with a floppy neck but with a heartbeat should be revived using a heat lamp and stomach tube. A vet can demonstrate how to give a glucose injection to revive a weak lamb as well. A lamb with hypothermia does not shiver.

A small lamb wearing a pullover made from the sleeve of an old jumper.

Conscious but unable to stand or feed If the lamb cannot feed from the ewe or a bottle, revive it using a heat lamp. Then feed from a bottle or stomach tube once warmed, preferably using the ewe's own colostrum.

Cold or lukewarm mouth If the lamb is cold and has a rectal temperature slightly below 39°C (102°F) but is lively and strong enough to cry and walk about, consider warming it with a heat lamp next to the mother (being careful to protect against fire risk and electrocution). Help the lamb to suckle from the mother or milk the ewe's udder into a bottle for the lamb.

CARE FOR THE NEWBORN LAMB CHECKLIST

- Spray iodine onto the navel shortly after birth to prevent bacteria entering the lamb's abdomen.
- Check the ewe has colostrum available by milking some from both her teats.
- Is the lamb attempting to get up?
- Does the lamb have any obvious abnormalities?
- Does the lamb have a sucking reflex when you put your finger in its mouth?
- Is the ewe bonding with her lambs and licking them clean?
- Ensure the lamb receives adequate colostrum in the first hour after birth (200ml for a 5kg lamb).

As your experience grows, you will also be able to assess whether the lamb has sufficient size and fat cover to have the energy needed to suckle and dry off. If not, it may need to go under a heat lamp

ORPHAN LAMBS

Adoption and Fostering

A shepherd's ambition is to see each ewe turned out to grass in the spring with two healthy lambs trotting along behind her. Artificial powdered milk replacer is expensive, and lambs reared on it never thrive as well as lambs reared naturally by a ewe on mother's milk.

'Spare' lambs that find themselves without a suitable mother can sometimes be adopted by a ewe that has lost her own single lamb or has a healthy single but is in good enough condition and has enough milk to feed two lambs. There are various techniques farmers use to adopt lambs onto other ewes; the wet fostering method described below is one of the most effective.

Ewes identify their own newborn lambs mostly by smell. If you can make the orphan lamb smell like the foster mum's own lamb, there is a chance she will feed it and look after the lamb as if it is one of her own.

Wet Fostering

For wet fostering to work, you need to be there when a single-bearing ewe is giving birth. Having an assistant to hold the orphan lamb can be a great help; otherwise have it nearby in a box.

1. Move the ewe to a small, secure pen and wait until she is ready to deliver the lamb.
2. When in the late stage of labour and something of the lamb is showing at the opening of the vulva, lie the ewe on her side (have an assistant to hold her head up if possible, but don't let her meet the spare orphan lamb just yet).
3. Assuming the unborn lamb is in a normal presentation, help the ewe deliver the lamb by pulling the lamb's feet alternately.
4. As you pull the lamb out, be ready with a bucket to catch as much of the waters

As the lamb is pulled out, use a bucket or large bowl to catch the birth waters from the ewe; this can be used to bathe an orphan lamb that you want to be adopted.

and slime that will follow the delivery as possible. Keep the ewe lying down. Before showing her the newborn lamb, wet the orphan lamb with as much of the foster mum's birth fluid as possible, particularly on the head and bottom. Rub the two lambs together to transfer more of the fresh slime to the orphan lamb.

5. Present the orphan lamb to the mother by lying it on the straw by her face. Keep the ewe lying down and give her five to ten minutes to bond with the orphan lamb before placing her own lamb for her to lick.
6. While she is meeting the orphan lamb, try to get the newborn lamb to suckle from the ewe's udder. The orphan lamb will usually be a day or two older than the newborn, so ensure it doesn't drink all the colostrum before the newborn has a chance to fill its own belly. Sometimes, tying the foster lamb's back and front legs together for ten minutes with rubber bands can help to make the lamb look like a newborn and prevent it getting up immediately to suckle.
7. Mark the foster lamb with a colour so it is easily distinguished. Keep an extra-close eye on ewes that have adopted lambs. Sometimes a ewe will appear to have bonded with her adopted lamb but after a few days you may notice she has rejected it.

Bottle feeding a triplet lamb. After one to two weeks post lambing, most ewes will struggle to provide adequate quantities of milk for triplets, so one lamb is taken away from the mother and bottle reared using powdered milk replacer.

Lambs that don't receive adequate natural colostrum at birth will always be more susceptible to disease. Colostrum contains immunoglobulins and forms the basis of the lamb's immune system.

Bottle Feeding

Powdered milk for lambs can be bought at agricultural merchants. Follow the instructions on the bag for quantities and mixing. Newborn lambs will need feeding every four to five hours. At seven to ten days old, this can be reduced to four feeds per day, equally spaced between 6am and 11pm. Depending on the size and strength of the lambs once they are two or three weeks old, they can be fed three times per day and given

CHECKING LAMBS IN THE FIELD

Once ewes and lambs have been turned out into the field, check them twice a day – morning and late afternoon. Walk quietly around the sheep, observing what they are doing. A healthy lamb receiving plenty of milk will either be sleeping contentedly or, when awake, will be full of beans, suckling strongly, jumping, running, playing. Observation both in the field and inside the lambing shed is important so you can spot problems early before ewes and lambs die or are taken by predators.

Lambs that are thriving will begin playing soon after birth.

In the first forty-eight hours after birth, a lamb that is being well fed by the mother will either be relaxing/sleeping, suckling or looking to play. A hungry or rejected lamb will be repeatedly trying to suckle, bleating and standing up for longer periods.

A Manx Loghtan ewe and her lambs looking very content.

'lamb creep' pellets. The lambs should always have access to hay and water and a clean, dry bedded area. Placing a small bale of hay or straw in the pen with the lambs gives them some comfort as they can lie against it and jump on it!

HUNGRY LAMBS

Standing with its back arched and constantly bleating, a hungry lamb may have been rejected by the mother or the mother may not be producing enough milk. Lambs receiving plenty of milk will be extremely hardy and full of vigour. Lambs that are short of milk can become cold and vulnerable to illness and predation very quickly.

A dirty patch on the back of the lamb's neck can be another indication of an underfed lamb. This happens when the lamb, rejected by its mother, tries to feed from her, or from different ewes, from behind at the same time another lamb is suckling.

When you pick up a lamb with a hand under the ribcage you can use the other hand to feel if it is full or empty.

MOVING A EWE WITH LAMBS

Pick up the lamb or lambs with a hand around the lamb's chest. Newborn lambs will be wet,

A ewe that has lambed for the first time may not follow easily but older, more experienced ewes will. If she runs away, put the lamb on the ground. Once the ewe locates her baby again, she will usually come back. The ewe follows the scent and sound of the lamb more than the sight of it so try to keep it close enough for her to smell.

slippery, and easy to drop in the first twenty-four hours after birth. A safe way to carry them is to hold the front legs in your hand just below the knee joint, with a finger in between the legs to prevent the lamb's knees being squeezed together. If you are working alone, you can confidently carry a set of twins like this.

Walk backwards and hold the lambs close to the mother's face and she will usually follow. Carrying lambs in this age-old shepherd's way may look awkward, but it does not hurt the lambs or stress the mother and is the safest way to carry two newborn lambs at once. It also means the lambs can be held in front of the ewe's face without the shepherd having to bend over, a real advantage if moving the ewe over any kind of distance.

An experienced ewe may follow the carried lamb easily and also remember her way to the field.

With a ewe that has sprightly week-old lambs at foot there is usually no need to pick up the lambs unless they are sick or hungry. The ewe can be ushered along with her offspring gambolling along on foot. A healthy lamb that is suckling well from the mother and getting plenty of milk changes rapidly by the first week of age. The lamb will become very difficult to catch so it is better to shepherd the mother and lamb into a pen if needs be.

A Portland ewe and her lamb. (picture credit Hannah Watson)

CHAPTER 14

COMMON HEALTH PROBLEMS AT LAMBING TIME

MASTITIS

Ewes with mastitis will often look dull and lethargic and may stay lying down when the rest of the flock are up and about grazing. A ewe with mastitis cannot produce milk for her lambs, so they may look hungry and will be trying to suckle frequently. The infected udder will be hard and swollen and there will be no milk. Early diagnosis is important.

The ewe may still be able to feed a lamb from her other teat but if she has twins, one lamb will have to be taken away and bottle reared. Talk to your vet about how to treat mastitis. Mild cases could be treated with natural remedies, poultices and lots of care but correct diagnosis is still important. Some of the antibiotic treatments used for mastitis in sheep can only be administered by a vet.

A ewe that has had mastitis will need to be removed from the breeding flock the following tupping season as she will no longer produce enough milk for her lambs.

Prevention of Mastitis

Mastitis is a bacterial infection inside the tissues of the ewe's udder. The causes are not always apparent, but with housed ewes, the risk of mastitis can be reduced by keeping the bedding clean and dry.

Cubicle lime (for dairy cows) can be spread around wetter areas in the shed before fresh

Keep clean and dry bedding in lambing pens and housed areas in order to reduce the incidence of mastitis in the flock.

straw is added on top. Ewes should not be expected to lie down in dirty and wet lambing pens. Muck the pens out between each set of ewe and lambs and provide fresh straw. Cleanliness also helps to reduce the incidence of bacterial infections developing in young lambs.

A cold wind and frosty or cold and wet conditions can give the ewes 'black bag', a severe and acute form of mastitis that causes blackening of the skin on the udder. In extreme cases, black bag will lead to rupturing of the outer wall of the udder and the tissues inside becoming dead and necrotic. Eventually the ewe will end up with a big hole in her udder and the insides falling

Housed ewes are more prone to mastitis but outdoor ewes that have freshly lambed can be at risk in very cold weather. Outdoor-kept ewes will still need clean areas of grazed grass on which to lie down. If a field is overstocked and muddy, the ewes will be more prone to mastitis and foot trouble.

out but she may die of septicaemia before this happens. Out in the field, hedges and areas of shelter from the wind are important to protect the ewes from this. Hill sheep breeds that have not had their tails docked will be less prone to this kind of mastitis.

Lameness and Mastitis

Lameness is linked to mastitis in some cases, as lame ewes spend more time lying down, making the udder dirty and 'clammy'. Treating lame ewes promptly and separating them from the flock helps improve their welfare and cuts the risk of mastitis infection. If possible, footbath the ewes using zinc sulphate solution before they are brought in for housing.

LISTERIA INFECTION IN EWES

Listeria can be a problem in housed ewes being fed on haylage or silage. Listeria is a bacterial infection and is usually caused by soil contamination in bales of haylage (one of the reasons why farmers do not like moles in their hay fields). Symptoms of listeria infection are sudden onset of black-looking

TAKING A EWE'S TEMPERATURE

If a ewe seems off colour, lethargic or not eating as well as the others, then taking her temperature can help diagnose an infection early. You will need a probe-type digital thermometer and sterilising wipes. Move the ewe into a small pen and put the thermometer into the ewe's rectum. If you have a helper to hold the ewe under the chin, it makes the task a lot easier.

The normal temperature range for a breeding ewe is 39–40°C (102–104°F). A temperature above 40°C (104°F) indicates an infection, and the sheep should be examined by a vet.

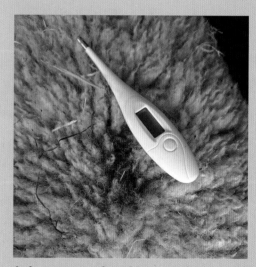

A thermometer for taking a ewe's temperature is very useful at lambing time.

diarrhoea, lethargy, not eating and standing in a corner staring into space.

A sheep with listeria infection cannot turn its head round to the left and may lose its balance when walking. Urgent veterinary attention should be sought in such cases, as the sheep will require a course of antibiotics if it is to survive.

TWIN LAMB DISEASE

A metabolic disorder of sheep carrying multiple lambs, twin lamb disease (pregnancy toxaemia) causes ewes to stop eating, become lethargic and usually too weak to stand. Early symptoms are failing to eat when the sheep are being fed, teeth grinding, twitching, standing and not responding to stimuli. It is easy to confuse twin lamb for other problems such as listeriosis, pneumonia and hypocalcaemia. Taking the ewe's temperature can help decide if the symptoms are caused by a disease pathogen or not. Seek veterinary advice if unsure, as the correct treatment will save the life of the ewe and her lambs.

Twin lamb disease occurs when the ewe is no longer receiving sufficient energy to support her and her lambs, leading to a drop in her blood glucose. The main risk period is around three to four weeks pre-lambing. It can occur in ewes that are too fat or too thin and can happen because the combined energy intake in forage and concentrate feed is not sufficient. In late pregnancy, it is important to have good-quality forage, as the quantity a ewe can physically consume reduces. She should have access to good, leafy, palatable hay or silage made in early or midsummer, as well as daily concentrate. A common diet fed to ewes in the last six weeks of pregnancy and a month after lambing is 500g (18oz) of concentrate per head per day split between two feeds.

Twin lamb disease can also be caused if a ewe is housed with more dominant types and she is pushed out at feeding time. This is more likely to happen if there is not enough feeding space for all the sheep to eat together. Other possible causes could be a sudden change in diet, or lameness.

Treatment

Treatment is usually by an injection of calcium and magnesium under the skin plus a drench of high-energy solution with electrolytes. A vet will often give a multivitamin injection as well. If dealt with early enough, the disease usually responds well to treatment, and the ewe will recover with some extra care over the following few days.

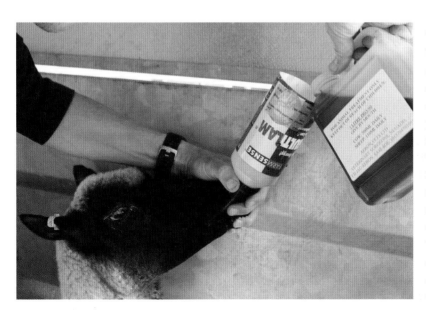

Treating twin lamb disease by drenching the ewe with a high-energy solution containing electrolytes. In an emergency, if a specialist preparation from a vet or farmers' merchants is not to hand, use a homemade remedy of four tablespoons of black treacle (or golden syrup) and a teaspoonful of marmite dissolved in warm water. Sports gels could also provide sugar and electrolytes to the ewe quickly.

HYPOCALCAEMIA

Ewes feeding two fast growing lambs can be vulnerable to hypocalcaemia (milk fever) after they have been turned out to grass in spring, especially if they are short of grazing, have been left without food for more than twelve hours or there is a sudden change to colder weather. Sheep fed straight cereals over a number of weeks with no added calcium can also be vulnerable, so adding finely ground limestone (calcium carbonate) to their feed can help to avoid milk fever.

A sudden drop in the amount of calcium in the bloodstream causes the affected sheep to stagger, collapse, fall over or have the neck turned back. If noticed in time, the condition can be cured by an injection of calcium solution into the bloodstream but sudden death can often be the first sign of milk fever.

COMMON INFECTIONS OF LAMBS

Joint Ill
Lambs that are limping may have 'joint ill', a bacterial infection contracted through the navel at birth. Consult your vet if concerned. The infection usually affects the joints of the front legs, causing them to swell and the lamb to hobble. Joint ill can be avoided by good ventilation in housing, using plenty of fresh bedding and keeping lambing pens clean. The navel of lambs should be sprayed with iodine as soon after birth as possible.

Watery Mouth
An *E. coli* infection in lambs up to about five days old is known as watery mouth because the main symptoms are a watery, wet mouth all the time, possibly a high temperature, sometimes diarrhoea and ceasing to suckle from the ewe. The infection can range from mild to severe. Prompt treatment with antibiotics prescribed by a vet is important.

Prevention of watery mouth in lambs is via good hygiene in the lambing shed by keeping the bedding in pens and group housing areas clean and dry. Putting lime and fresh straw around feeding and drinking areas will help to prevent bacterial infections in ewes and lambs.

PREDATORS

In Britain, lambs are vulnerable to predators up to about fourteen days of age, and will be at greater risk in fields far away from the house or near woodland.

Foxes
A good, attentive mother will do her best to fight off a fox or badger. Defending twin lambs, especially if one is weak, may be a losing battle, however, particularly for a young ewe. If several lambs under three or four days old are being killed by foxes, it is a good idea to delay turning the lambs out into the field. Unfortunately, if a fox is successful in taking a lamb, it is likely to return.

If this happens, try to identify where the fox is coming into the sheep field. Foxes are habitual and often follow the same runs each night. Look for tracks in the grass and for hairs on the fence wire where a fox might push underneath. One year I lost several lambs to a fox in the same 1.5ha (4-acre) field. I put up an electric fence on two sides of the field where it adjoined a woodland. The fence had two wires running close to the ground, which seemed to be enough to deter the fox. Urban foxes can be more of a problem than rural ones, as they are bolder and will attack in the daylight. For outdoor-lambing ewes, removing any ewe's afterbirth from the field could help to deter predators.

There are people in the countryside who offer to shoot foxes using a high-powered torch and a rifle (they do it for sport). If you allow this, check they are experienced and ask the local police if they have a current firearms licence. This is very important as stray rifle bullets can travel over a mile in the air. Anyone shooting foxes must have the landowner's permission.

Birds

Crows can be a problem at lambing time. A ewe lambing in the field or getting stuck on her back will be vulnerable to attack by crows, who will peck the eyes out of a stricken ewe or lamb. Crows can be trapped with Larsen traps if they are a persistent problem.

Birds of prey will usually only eat lambs or ewes that have already died from another cause, except eagles in Scotland, which will take lambs to eat. If you see large bird of prey or crow activity around the sheep field, check the sheep more closely, as the birds will know if an animal is weak from disease or stuck in some way. Killing birds of prey is illegal because many species were driven close to extinction in the twentieth century.

CHAPTER 15

SHEEP LAMENESS

Research has shown that it is not uncommon for around 8 per cent of sheep in a flock to be lame at any one time.

A ewe with a lame front foot. Lame sheep are difficult to catch and treat in the field, making it necessary to run the whole group into a pen or yard.

'Footrot is a contagious, acute or chronic dermatitis involving the hoof and underlying tissues (Bulgin, 1986). It is the leading cause of lameness in sheep.' sciencedirect.com

Unfortunately, lameness is common among sheep in the UK, and it is not unusual to see within any flock a few sheep holding a foot off the ground, grazing on their knees or walking with a limp. A 2011 report by the Farm Animal Welfare Council (FAWC) describes sheep lameness as a common problem worldwide. In 1994, sheep farmers in the UK estimated that 8.4 per cent of their flocks were showing signs of lameness at any given time. The report notes that the incidence of lameness a decade later was estimated to be 10 per cent in UK flocks. (The report brings together an overview of the research and treatments and includes references to further information https://assets.publishing.service. gov.uk/government/uploads/system/uploads/ attachment_data/file/325039/FAWC_opinion_ on_sheep_lameness.pdf.)

Sheep that are lame for one week or longer lose body condition and become debilitated and less productive.

FOOTROT AND SCALD

There are two main types of bacteria that cause lameness in sheep: *Dichelobacter nodosus*, known colloquially as footrot, and *Fusobacterium necrophorum*, or foot scald. When combined, these two types of bacteria can cause severe lameness in susceptible sheep.

F. necrophorum is present in the soil and manure. It causes a dermatitis that erodes the skin (causing death of living tissue) and appears as a whitening and/or reddening of the skin between the hooves and where the hoof wall joins the skin of the leg.

Vulnerability to foot infection is influenced by genetics and environmental factors. Scald

An example of a severe infection of foot scald. It has spread from the gap in the centre of the hoof outwards to the soles of each cleat.

is worse in mild, wet conditions and when the grass is longer than usual and growing rapidly, as it is in the spring and autumn.

Sheep suffering from a clinical infection with scald (that is, exhibiting lameness) will spread the bacteria to other sheep, especially in places where they gather, such as watering spots and around mineral licks. An animal with a sore foot due to an outbreak of scald can also pick up secondary bacterial foot infections, which can embed deeper inside the tissues of the foot.

A more common cause of severe lameness, however, is footrot, which presents with bloody open sores on the foot and an unpleasant smell. The shell of the hoof often peels off, leaving sore flesh underneath. Footrot is also contagious and spreads from sheep to sheep. Unlike scald, the footrot bacteria cannot survive more than around two weeks on pasture, so it can be eliminated with a good treatment and control regimen.

OTHER FOOT PROBLEMS

CODD
Contagious ovine digital dermatitis (CODD) is another common reason of serious lameness in UK sheep flocks and is also caused by bacteria, though it is little understood. Like footrot, CODD is contagious, spreads from farm to farm and can be introduced to a flock by bought-in animals, even if they show no signs of lameness. CODD causes abscesses along the line of the coronary band - the line where the horn of the hoof joins the leg. In severe infections, the whole hard casing of the foot (the horn) can peel off and the farmer may decide to cull the affected sheep on welfare grounds. Treatment is with a course of injectable antibiotics and a recovery period indoors will be required of two to four weeks.

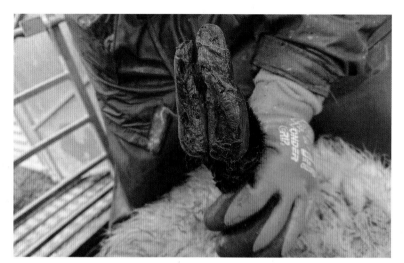

Shelly hoof is a genetic defect. The horn material at the sides of the foot becomes hollow, allowing dirt, stones and bacteria to accumulate. Here the shelly hoof is extended into a hollow, overgrown toe as well.

Shelly Hoof

Also known as 'white line disease', shelly hoof is a problem that sheep inherit from their parents, with some breeds being more affected than others. The problem occurs when the horn on the outside of the foot becomes separated from the sole of the foot. A cavity opens and this fills with dirt, muck and grit in the field. The cavity creates the perfect moist anaerobic environment (without oxygen) for scald bacteria to reside in.

The problem with shelly hoof is that footbathing is ineffective at treating it unless the loose horn on the side of the foot is trimmed away. Do this very carefully, being careful not to cut into the toe or the softer white/pink foot layers (the quick) underneath. Footbath again after trimming.

Foot Abscesses

Sheep can suffer from an abscess, usually at the toe of the foot. Bright red bubble-like abscesses can be caused by foot rot bacteria. They are difficult to treat. A foot abscess requires that the pus is drained from the foot and the sheep is treated with an injectable antibiotic. A ram that is prone to either problem should not be kept for breeding, as susceptibility to abscesses can be inherited.

FOOT OR LEG INJURY

Keep in mind that lameness is not always due to foot infection; it could be because of an injury to a leg or because a foreign body like a thorn or piece of stick, metal or ball of mud is stuck in the foot. Close inspection of the lame sheep is required before deciding on

A toe abscess, also known as 'strawberry footrot', is spread by bacteria from sheep to sheep. Isolate affected sheep until after treatment and continue with regular footbathing the rest of the flock, moving to clean pasture afterwards. It can be helped with trimming, rest in a clean dry pen and antibiotics, but is likely to recur, so may be a cause to cull the affected sheep from the flock.

treatment. Leg injuries are common if sheep break loose or are chased by a dog. An injured sheep should be kept in a dry pen with access to hay and water until the leg heals. Vets can prescribe painkillers and anti-inflammatory drugs if necessary.

TREATMENT AND CONTROL STRATEGIES

Biosecurity and Quarantine

The prevalence of footrot in the national flock highlights the importance of treating and quarantining any new animals brought onto your farm and keeping secure boundaries to prevent neighbouring sheep from straying onto your land.

If you have worked hard to eradicate footrot and CODD in your own flock then quarantining new stock and improving biosecurity with good double fencing between neighbouring sheep farms should be a priority. The quarantine period needs to be at least twenty-one days, as clinical signs of footrot in sheep can take two to three weeks to emerge. Standing sheep in a medicated footbath twice during the quarantine period should eliminate the footrot pathogen from the feet of sheep brought in.

Footbathing

Scald and footrot can be controlled by regularly footbathing sheep. At the same time, lame sheep should be separated from the group and treated with antiseptic sprays or injectable antibiotics until they are better. Footbathing is useful for controlling the spread of the disease, but on its own it is not usually a sufficient treatment for clinical cases of footrot.

The key to successful footbathing treatment is to have a clean (not muddy or mucky) area where the sheep can stand

Regular footbathing every six to eight weeks with zinc sulphate solution can help to keep infections of foot scald under control, as can separating affected individuals from the flock as soon as possible.

before and after the footbath, as it will work much better if the sheep's feet are clean. An effective way to clean the feet is to first walk sheep through a footbath holding just water, then into a second footbath holding the medicated solution.

For treating an outbreak of footrot the sheep should ideally be footbathed twice in a week.

If a set-up of hurdles and a sheep footbath is not available, individual sheep can be turned over onto their backs. The feet can then be cleaned with a small bucket of warm soapy water and a brush, and treatment applied to the clean foot.

Footbath Ingredients

There are a couple of choices when it comes to the active ingredient in the footbath.

Zinc sulphate or copper sulphate These are both salts that can be bought in 20kg (44lb) bags from agricultural merchants and dissolved in water. Effective against scald, they can also be used for treating non-severe footrot if the sores are not too deeply embedded in the foot. Cases of footrot will require at least two treatments in a week and keeping the sheep on clean dry bedding in between if possible.

Formalin This is a powerful antibacterial chemical; it is a derivative of formaldehyde and a known carcinogen. It can be bought in 20-litre drums from farmers' merchants. It is unpleasant to work with due to its odour, and the health risk to the operator requires that protective clothing be worn. The advantage of formalin over zinc sulphate is that sheep only need to walk through it, not stand in it for a period of minutes, therefore it saves time in larger flocks of sheep.

Adding old wool to the footbath or pieces of foam sponge can help to calm the sheep's fear of water and reduce splashing.

Foot Sprays

Foot sprays for sheep are distributed by vets and agricultural merchants. There are two types – antibiotic, and antiseptic. The former must be prescribed by a vet. Both types of spray need to be used on clean feet to be effective, and the animal should stand on a clean, dry surface afterwards.

A footbath set-up of fixed hurdles with metal sheeting on the sides and a gate at each end is a useful addition to the sheep-handling area.

Injectable Antibiotics

Where an infection has embedded itself deeper inside the foot, bathing and foot spraying alone will not be sufficient. In this situation, the sheep will require a course of injectable antibiotics to clear up the infection. Consult your vet about prescribing a suitable antibiotic. Treatments must be recorded in your flock medicine book, including the ID of the sheep, the date, medicine name, quantity, batch number, expiry date and the withdrawal period of the medicine.

RECORD KEEPING

It is a legal requirement to keep records of veterinary drugs used on your holding. Good records can also aid you in the management of the flock and help improve the quality of the flock over time. For example, if you know of recurring health problems in the flock and can identify animals that are affected more than others you can keep breeding replacements and rams from only the healthiest most robust sheep in the flock.

Antibiotic Resistance

Antibiotics should never be used as a 'just in case' preventative measure or to mask bad management of the flock. Sheep that have received a course of two or three injections of antibiotics will recover from lameness in the short term but remain susceptible to reinfection and may require repeated treatment. Strategies for controlling sheep lameness should only include antibiotics as a treatment of last resort where treatment is needed to improve an animal's welfare. Antibiotics do not build immunity and can even decrease immunity to a pathogen in the longer term; meanwhile, the target bacteria can become immune to the drugs with incorrect or overuse.

Since the 1970s, the proliferation of animal farming on an industrial scale led to overuse of antibiotics in livestock farming, especially in the US, Europe and Australia. The non-therapeutic use of antibiotics saw them being used widely over a long period as growth promoters in cattle, sheep, pig and chicken farming. The anti-microbial drugs were also used prophylactically, to keep animals 'healthy' when kept in unsanitary, stressful and overcrowded conditions. Micro-organisms exposed to a low dose of antibiotic over a prolonged period become resistant to the drugs used, which were routinely added to feed and water. This practice was fully banned in the EU in 2006, but the damage was largely already done. (An example of a drug-resistant strain of bacteria is MRSA, which is found in both farm livestock and humans. This is one of the most serious and deadly hospital-acquired infections worldwide.)

The farming industry is now more aware of the dangers that drug resistant pathogens pose to human and animal health, and education campaigns for farmers and vets encourage the responsible use of antibiotics.

Foot Trimming

Foot trimming can be a controversial subject. Traditionally, shepherds would trim the feet of their sheep at least two or three times per year. This may not have been the case for sheep on the harder, rockier ground of the hills, but for larger lowland breeds kept on soft, fertile fields of grass all year round, foot trimming was seen as an essential preventative measure against lameness.

A labour-intensive and back-breaking job, it at least gave the shepherd the opportunity to inspect every foot and identify individuals with poor feet that should not be kept for future breeding. With regular foot trimming, foot infections can be identified and treated.

The problem with routine foot trimming is that in many cases the stockperson working on the sheep would trim away too much of the horn, exposing the softer layers of tissue and opening them to damage from stones and abrasions and more prone to infection. In many areas there was even a barbaric belief that the foot should be trimmed until it started bleeding. This was thought to release infections but instead it opened the foot to infection, causing severe pain and abscesses and granulomas (scar tissue).

This image shows where the toe should end (the shorter side) and how the other is very overgrown. Consider culling sheep with feet like this or at least not keeping their offspring for breeding. If an overgrown hoof is impeding walking, carefully trim back the overgrown horn; avoid cutting into the tissue of the sole or the toe, as there is an important blood vessel at the end of the toe.

Today, the current advice by vets and farm trainers is **not** to practise routine foot trimming on all members of the flock. Scientific studies have shown better outcomes when only lame sheep are targeted for foot trimming and sheep that are not lame are left well alone.

When to Trim

If you feel you must trim a sheep's feet because the horn is overgrown and flaking off, be mindful to trim the foot sparingly. Remember the horn on the outside of the foot protects the softer tissues inside, and only cut away loose horn growth that is coming away from the foot. The outside horn of the foot should be left a few millimetres proud of the foot sole so that the weight is carried more on the outside and the sole has some protection. Do not cut into the flat pad at the base of the foot or into or across the toe area.

Summary of Management Practices to Reduce Lameness in the Flock

- Quarantine and footbath any new animals brought into the flock for at least three weeks.
- Separate lame sheep from the rest of the flock as soon as possible to prevent lameness spreading.
- Use footbathing as a regular whole flock treatment to manage scald and identify sheep suffering from lameness.
- Clean the sheep's feet before treatment so you can see what the problem is and to allow sprays to be effective.
- Turn over and inspect any sheep limping after footbathing. If an infection looks to be only on the surface, spray the foot with

antiseptic or antibiotic spray (prescribed by a vet).

- If a sheep is severely lame (not bearing weight on the affected foot), has an oozing open wound or deep cavities that are infected, treat it with a course of injectable antibiotic after consultation with the vet.
- Use spray paint or crayon to mark the wool at the top of the leg belonging to the treated foot. This allows you to monitor if the same foot remains a problem. Note the tag number of the affected sheep in your flock records.
- Have a dry 'hospital pen' with hay and straw available all year round so that lame or ill sheep can be treated away from the flock.
- Do not return lame sheep to the rest of the flock until they have fully recovered.
- Practising rotational grazing, where the sheep are moved to fresh pasture every two to three weeks, can help control contagious foot problems as well as internal parasites, and will boost the sheep's growth and vitality.
- Do not keep breeding replacements from ewes or rams that have a history of foot trouble and cull from the flock any sheep that have persistent foot trouble and poorly shaped feet.

Identifying and treating lame sheep early can help avoid the need for antibiotics.

CHAPTER 16

INTERNAL PARASITES

Chicory leaves are very palatable to sheep, and studies have shown that its high tannin content helps to reduce sheep worm burdens by suppressing the development of the parasites.

Along with lameness and flystrike (*see* Chapter 17), gut worms are a major source of welfare problems in sheep. As sheep graze, they pick up parasitic worm larvae from the pasture. The worms damage the lining of the sheep's stomach and can cause weight loss, diarrhoea (scours) and, in some cases, death. The parasites reproduce inside the sheep and the parasite eggs are excreted in the animal's dung. When conditions are favourable to the parasites (warm, moist and

Drenching lambs at weaning time before moving them to clean grazing. Decisions on when or if a worming product (anthelmintic) should be administered to the sheep need to be taken carefully. Timing and management of grazing is crucial. Making a flock health plan and gathering evidence of infection level can aid decision making.

sheep over-grazing a certain area), the gut worms can increase exponentially in a short space of time, leading to clinical disease in the sheep.

SIGNS OF WORM INFESTATION

Sheep carrying a high worm burden will often (though not always) be scouring (suffering diarrhoea).

Sudden onset of scours in lambs can indicate a sudden rise in worm numbers, though sometimes it can be caused by a change in diet or pasture quality.

Scouring (dirty wool around the back end) is a sign of gut worm infection.

Some sheep in the flock will have lower natural resistance to internal parasites than others, but the whole flock may need treatment at certain times. Diagnosis is best carried out by having a dung sample analysed through a veterinary lab.

This does not mean you should constantly dose your sheep with wormer medication. In the first instance, it is better to try to control the prevalence of internal parasites through the management of your land and the flock.

NATURAL RESISTANCE

Sheep flocks always carry a burden of parasites with them. When parasite numbers are low and the sheep have natural resistance, an infection with gut parasites will not cause clinical disease. Natural resistance is determined by age, breed, the level of inherited genetic resistance but also by how a sheep is reared from a lamb. If a sheep has natural resistance to worms, its immune system will suppress the development of the larvae inside its intestines. Sheep with low natural resistance to worms spread more worm eggs onto the pasture than healthy sheep, so they should be identified and removed from the flock.

Some sheep have little natural resistance to internal parasites, and even a low level of infection can make them ill or drastically stunt their growth as lambs. Lambs who do not receive adequate maternal colostrum within an hour of birth are much more prone to gut worms making them ill throughout their lives, with even low levels of infection causing symptoms like scour, weight loss and damage to the lining of the gut. Lambs reared on artificial colostrum and bottle-fed with powdered milk are more prone to disease, as they do not receive the antibodies present in natural ewe's milk.

Well-fed, healthy sheep that have inherited immunity through colostrum and ewe's milk can cope with a degree of exposure to internal parasites and remain healthy.

DIRTY GRAZING

Pasture that is continually grazed by sheep carries a higher worm burden and is viewed as 'dirty' pasture for sheep. Fields that are heavily grazed with sheep over summer and rested in winter still pose a risk, as worm eggs will survive on the pasture over winter and hatch the following spring, infecting the flock again.

On land where sheep graze every year, lambs are at risk from worm infections developing into clinical disease. They have lower natural resistance than ewes and, as the flock grazes from springtime into summer, the worms reproduce inside the sheep and more eggs are excreted. With each completed worm reproduction cycle, the number of parasites builds up, so that by early summer there is a risk of lambs losing weight and becoming ill because of high worm infection levels.

The first warm weather in the spring is also a risk period for lambs, as worm eggs that have lain dormant over winter all hatch into parasitic larvae at the same time.

CLEAN GRAZING

Clean grazing is pasture that has not had sheep grazing it for at least twelve months. This is hard to achieve on most sheep holdings unless there is a crop rotation including arable or vegetable crops. If the land can be kept only for hay production, or grazing by cattle or horses for a full calendar year, it will be clean for sheep afterwards. If you are lucky enough to start a new flock on a piece of land that has not had sheep or goats grazing for a year or more, it is advisable to drench the sheep twenty-four hours before turning them out onto it. This will ensure you are not introducing worm eggs onto the clean pasture.

On many holdings, it is simply not possible to have a full clean grazing system in place, but there are management techniques to help keep worm burdens manageable.

LAND MANAGEMENT TO REDUCE WORMS

As discussed in Chapter 2, natural ways of controlling sheep parasites are possible, and they not only benefit the health of the sheep but increase production from the land.

Lower Stocking Rates

Having lower stocking rates (the number of sheep kept per hectare) can lower the overall worm burden on the sheep.

Rotational Grazing

Even with low stocking rates, internal parasites can still be a problem unless sections of the pasture are periodically rested from grazing and the sheep rotated around. If left on a large area, a small number of sheep will have their favourite bits of the field and graze them more frequently, thus building up numbers of parasites in those areas.

Rotational grazing by moving the sheep regularly and resting pasture can help break up the parasites' life cycles. If the sheep are moved on regularly, the worm eggs excreted onto the pasture will hatch into larvae but not find a host in which to reproduce.

A low stocking rate (number of sheep kept per hectare) gives sheep more fresh forage to choose from. Sheep in good body condition are better able to resist parasite infections.

Rotational grazing not only increases pasture production and lamb growth rates but can aid in keeping worm burdens on the sheep manageable.

Mixed Grazing

Mixed grazing with other species, such as cattle and horses, again helps to disrupt the parasites' life cycles. Worm numbers are controlled as the cows or horses ingest and destroy the worms that affect the sheep with no detrimental effect to them. This does not work with goats, however, as they are affected by some of the species of worm that affect sheep.

Nutrition and Minerals

Providing sheep with enough quality pasture and mineral supplements and concentrate feed when needed helps to keep them in good

Mixed grazing can interrupt parasite life cycles.

condition. Sheep in good body condition can support their own immune systems better than sheep that are malnourished or deficient in certain minerals.

Herbs and Hedgerows

A diet that includes herbs in the pasture can help to suppress parasites when infections are at low levels. The high tannin content of chicory, for example, acts as a natural anthelmintic when consumed by sheep. The leaves of hedgerow plants also help provide a diversified diet that is higher in minerals and tannins than grass alone.

MONITORING WORM BURDENS IN THE FLOCK

In summer, the worm burden on a pasture can increase rapidly from one week to the next. The level of parasite challenge to the sheep can be monitored by faecal egg counting (FEC). This involves collecting fresh dung samples, looking at the faeces under a microscope and counting how many worm eggs there are per gramme of faeces. An FEC kit can be purchased for the job, but most vet practices will offer this service. A worm egg count below 250 eggs per gramme of faeces is not considered to cause sheep ill health. Above that level, and the sheep will need a control treatment with a worm drench or injectable wormer.

Monitor lamb development and growth rate; weighing lambs can be useful for this. If growth is checked, or lambs begin scouring, investigate for internal parasites and consider drenching.

Ask a vet to arrange a post mortem of any sheep that die of sudden natural causes. Sometimes death can be caused by internal parasites; for example, *Haemonchus*, or barber's pole worm, can cause sudden death in sheep. The worms attach to the lining of the gut, sucking the sheep's blood and rapidly causing anaemia and toxicity. If it is identified by a post mortem, the rest of the flock can be drenched with a suitable product to prevent further losses.

Tapeworm infection is visible to the naked eye by looking at the sheep dung – tapeworm segments look like large white discs in the faeces. Tapeworm does not normally cause illness in sheep but can cause weight loss, and is also transmissible to dogs and other animals. Not all drenches will clear tapeworm from the sheep so seek advice on an appropriate product to use.

Lungworms affect both sheep and cattle. Its presence is often evidenced through a persistent dry-sounding cough, especially when the animal has been walking fast or running. It can be identified by a blood test or dung sample examined in a lab. Try to find out if your land has a history of lungworm and liver fluke problems.

TREATMENT WITH ANTHELMINTIC DRENCHES

As far as possible, use the natural controls detailed above to manage intestinal parasites. When a drench is necessary, seek advice on which type to use. In the UK there are many different brands and products of sheep wormer but they all broadly fall into four main types:

- **White drenches** Albendazoles, benzimidazoles
- **Yellow drenches** Levamisole
- **Clear drenches** Ivermectins (has negative effects on soil life); injectable ivermectins can be used to control internal and external parasites
- **Orange drench** Zolvix, the only type on the market at present; it is costly but there is no known resistance at present

Rams and ram lambs being brought in from pasture to be given a dose of wormer (late summer). Dagging - clipping the dirty wool from their back ends - can be done at the same time.

Drench Resistance

It is important to be aware of the widespread problem of drench resistance as it can compromise the welfare of the sheep if the anthelmintic product used does not lower the worm count for the sheep.

Due to historic overuse of anthelmintics by sheep farmers, internal parasites have become resistant to the drugs used to control them. This means a lot of the products on the market today are ineffective against treating gastro-intestinal worm infections. The only way to know is to take dung samples from a set number of sheep and analyse them for a worm egg count. Then treat the sheep with a drench and carry out another worm egg count afterwards. If there is no significant reduction in worm eggs, this indicates the parasites have drench resistance. Your vet can assist with this process or recommend

a product to which there is no known resistance.

Quarantine Drench

Before allowing newly bought-in sheep access to your pasture, consider giving them a quarantine drench using a product to which there is no known resistance. This will help to avoid introducing drench-resistant internal parasites onto your farm.

LIVER FLUKES

Another parasite that can cause serious disease, liver flukes affect both sheep and cattle so mixed grazing is not a control strategy. Immature fluke larvae are ingested by sheep when grazing, then migrate into the liver to mature and reproduce. The flukes damage the liver and can cause anaemia, poor liver function, weight loss, a severe weakening of the animal and, in extreme cases, death. When sheep are taken to an abattoir the slaughterhouse staff have a responsibility to tell the producer if there are signs of fluke infestation in the liver.

Liver fluke is a problem on farms that have areas of soil that are waterlogged for most of the year, as part of the fluke's life cycle is in the body of a species of mud snail. Wet soil conditions are needed for the host snail to live in. This is why liver fluke is more common on farms in the western and northern areas of Britain, where rainfall is higher. Soil type is also a factor in the prevalence of fluke. Heavy, silt, clay loams that do not drain freely will hold onto

moisture all year round. The dark-green soft rushes seen growing prolifically in western regions indicate where ground conditions are suitable for liver fluke to survive.

Treatment

If wet areas account for only a small proportion of your overall grazing land, the simplest thing might be to fence those areas off so sheep cannot graze there.

Flukes are slower maturing in summer than gastro-intestinal worms, so treatments are usually only necessary in autumn and winter. Monitor the sheep for signs of weight loss and anaemia. Signs of severe anaemia are a yellowing of the mucous membranes around the eyes and yellowing inside the gums. However, an animal with obvious

Having given birth, the immunity of ewes to internal parasites can temporarily wane, leading to greater numbers of worm eggs being excreted. This is known as the periparturient rise. Lambing time is a good period to test for worm eggs and treat the ewes if necessary.

anaemia would already have significant liver damage. It is better to be vigilant for fluke and look for the presence of it in October or November by taking fresh dung samples to a vet for analysis. Vets can also diagnose fluke infestation by a blood test. If fluke is identified as a problem on your farm, develop a control plan in conjunction with your vet.

Over-the-counter drenches are available for the treatment of fluke but only use them if there is an identified diagnosed problem.

SUMMARY: MAKING A PLAN TO CONTROL INTERNAL PARASITES

- Understand the problem through monitoring and accurate diagnosis.
- Implement good management of grazing land to control parasites. If possible, design a system of rotational grazing, mixed grazing or crop rotation to disrupt parasite life cycles.
- Keep a watchful eye on lamb growth so weight loss can be spotted quickly.
- Develop a health plan in conjunction with a vet that includes monitoring for parasites and carefully timed treatments when required.
- Quarantine and drench bought-in sheep to avoid introducing resistant strains of parasite onto the farm.
- Be especially vigilant during high-risk periods, particularly in spring and early summer, when warm moist weather can cause a mass hatch of worm eggs on the pasture. This will be a larger problem if the land had continuous sheep grazing the summer before.

CHAPTER 17

EXTERNAL PARASITES

Sheep in full fleece in early summer are susceptible to flystrike, so shepherds need to be vigilant and check every day for symptoms of skin irritation in the sheep's behaviour.

FLYSTRIKE

All sheep keepers should be aware that in summer and autumn sheep are at grave risk from flystrike, which is caused by the common greenbottle fly, or blowfly. The female flies are attracted to dirty wool with urine and faeces on it or areas where the skin might be damaged. They lay eggs on the sheep that hatch into maggots within twelve hours. The maggots begin digesting the surface layers of the skin, causing a wet appearance and sores. Once a sheep is 'struck', more flies are attracted by the smell and the problem rapidly becomes worse.

Sheep should be inspected every day in the summer and early autumn months for signs of flystrike. Early infection with maggots is not visible from a distance but a struck sheep will exhibit behavioural signs such as kicking the affected area, nibbling the wool, and being restless and distracted while its flock mates are grazing or relaxing. You may be able to see a dirty area of wool from a distance, but not always. It's worth keeping in mind that sheep can get flystrike and maggots in a foot as well. This occurs when flies are attracted to the smell of footrot. In severe cases of flystrike, a sheep will be so infested with maggots that it will cease struggling and lie down somewhere to wait for death by septicaemia.

Close inspection of a sheep with flystrike reveals a dirty-looking, moist area of wool. The wool usually lifts away from the skin easily, revealing a heaving mass of maggots. These can range from immature thread-like larvae to full-grown maggots in the worst cases.

Risk Factors

In the UK, flystrike usually starts to be a problem in late spring, although the timing varies according to geographical location and weather. Keep a look out for greenbottle

Discoloured wool with maggots underneath. Rapid diagnosis and treatment of flystrike is essential, otherwise the sheep will suffer greatly and die. This sheep was struck in late August, when the effectiveness of the first preventative chemical treatment used after shearing was waning.

Ewes and lambs in May before shearing time are at risk of flystrike. Sheep are most at risk in moist, warm conditions where there is little wind. In early summer, a sheltered, boggy grassland such as this would give a heightened risk of flystrike over a windy hilltop with short grass.

and bluebottle flies around fresh sheep dung. If you see them, the sheep will be at risk. Blowfly emerge for a new breeding season when the conditions are sufficiently warm and moist. Sheep in full fleece are at risk, especially if they have areas of soiled wool or are scouring and dirty around the back end. Unfortunately, it is those sheep already suffering from scours caused by gastro-intestinal worms who are most vulnerable to blowfly strike.

It's common to find maggots around the tail and back end of sheep but flystrike can occur anywhere on the back, shoulders and flanks as well.

Preventative Measures

Prevention is better than cure, as maggots on a sheep rapidly become a serious welfare problem. Keeping the sheep clean is important. Dagging the sheep (close trimming of wool around the tail) in early spring can help to keep them clean before shearing time. Dealing with any scouring sheep promptly will also help to prevent maggots.

Shearing in early summer removes the previous twelve months' worth of wool growth from the sheep. After shearing, the sheep are clean and for a few weeks are at much lower risk from blowfly.

In late summer and early autumn, however, the risk of flystrike increases again. Rain in August and September causes a rapid regrowth of grass. This can upset the digestion and make the sheep dirtier around their tails. Lambs will usually need dagging in late summer/early autumn to keep them clean.

Pour-on Chemical Controls

In Britain, chemicals designed to prevent blowfly strike in sheep are available from agricultural merchants. Some products give six to eight weeks' protection and some give sixteen. The longer-protection products are more expensive and have a longer meat withdrawal period, so are not always suitable to use on fattening lambs. Each agricultural merchant has a staff member trained to advise on animal medicines. Follow the instructions on the bottle and ensure you and your assistant have gloves, masks, waterproof trousers and coat to make sure none of the chemical can reach your skin or be inhaled. The chemical is applied using a special spray gun attached to the bottle with a tube. Each sheep receives a wide stripe of chemical applied from the shoulder to the tail and across the bottom.

When to apply the treatment will depend on the location of your farm. Ask local sheep farmers when they usually shear the sheep and when they start applying the pour-on chemical to prevent flystrike. The chemical is best applied two to three weeks after shearing. Shearing should not be delayed, as this will increase the risk of flystrike. In the UK, the National Animal Disease Information Service (NADIS) runs a useful website, nadis.org.uk, which give parasite forecasts for each region of the UK.

On my farm in mid-west Wales, the sheep are sheared during the first week in June (weather permitting). The lambs are separated from the ewes at shearing time and

A pour-on chemical is applied to the ewes' backs and around the tail and rump in mid-June, a couple of weeks after shearing.

a pour-on chemical that gives seven weeks' protection is applied to the wool along their backs and around the tail. After shearing, the lambs are reunited with the ewes before being re-treated at the end of July when they are being weaned. The chemical treatment will not be used on any lambs that are ready to sell for meat at this time because of the statutory meat withdrawal period.

The ewes, meanwhile, are gathered in again two to three weeks after shearing and a pour-on chemical is applied to the new wool that gives them sixteen weeks' protection. This will last until the end of the risk period, which is usually some time in October.

Treatment

If you find or suspect a sheep has been struck by blowfly, catch the animal immediately. Part the wool to determine the extent of the damaged area of skin. Using (preferably electric) clippers, clip off all the wool from around the affected area. Follow the tracks of discoloured wool left by the maggots and make sure you remove all the maggots from the sheep.

Once clipping is finished, use a chemical such as 'spot on' deltamethrin or 'crovect' cypermethrin to kill any immature larvae that may still be on the sheep.

Clip all the wool from around the affected area.

The sheep pictured was sprayed with an antiseptic and a chemical to avoid flystrike and made a full recovery, although the wool growth in the affected area never recovered completely.

An advanced case of flystrike shown after wool clipping. Fortunately, the maggots did not break the skin and the ewe recovered. Ideally a sheep should be caught and inspected before the obvious wool discolouration gets to this stage.

If the sheep's skin has not been broken by the maggots, spray the affected area with an antiseptic spray and a pour on flystrike prevention chemical. The sheep will then be fine to rejoin the flock.

If the maggots have broken the skin and made open wounds, the sheep will be at high risk of secondary bacterial infection of the wounds. A vet should be consulted so that a course of antibiotics can be prescribed.

In the worst-case scenario where a sheep has not been treated and has been left for several days, the animal may have to be euthanased on welfare grounds. Lambs are particularly vulnerable to acute damage by maggots, as their skin is thinner.

LICE

Biting and chewing lice can be an irritant to sheep in winter, when they are in full fleece. If the sheep are observed constantly rubbing on fences and gates or trying to rub the flanks or neck with their back feet, investigate for signs of lice. Part the wool and look for brownish-orange specks. With a severe infestation, the sheep will spend more time rubbing than grazing, will have patches of loose wool and will lose weight.

Chemical products to treat sheep for lice infestation can be bought from most agricultural merchants. Observe the necessary withdrawal periods for animals producing meat or milk for human consumption. Two treatments covering the whole flock, with an interval of 3 to 4 weeks, are usually necessary to clear a lice problem.

SHEEP SCAB

Sheep scab is endemic in commercial sheep flocks in Britain, especially on the hills and commons. It is caused by a microscopic mite that lives on the skin and causes intense irritation to the sheep. In its initial stages, sheep scab can be mistaken for an infestation of lice, as the sheep rubs and chews its fleece, and kicks its flanks and neck, leaving dirty patches of wool.

Sheep scab develops into more severe disease than lice as the irritation and rubbing becomes worse, bare patches of skin develop and the skin becomes thickened, dry, red and scabby. Eventually the sheep will suffer severe stress, irritation, pain, weight loss and be likely to die of a secondary cause if left untreated.

Be vigilant for signs of sheep scab in your flock. If you suspect its presence, consult a vet as soon as possible for correct diagnosis and treatment. Good biosecurity, such as having double-fenced boundaries and quarantining bought-in stock, can help prevent your sheep catching scab.

The NADIS website is a useful resource for in depth information on sheep scab as well as all other diseases of sheep, as is the www. scops.org.uk website.

TICKS

In some areas of Britain, particularly on hills, heathland and rough grassland, ticks are a consistent problem. Ticks survive and reproduce in thick, damp undergrowth. Part of the parasitic tick's life cycle is on a sheep or other mammal, where it feeds on the blood. They usually attach to the haired areas of the sheep, such as the head, neck and legs, and cause intense irritation and itching. Ticks also carry bacteria, which can occasionally cause disease in sheep, other mammals and humans. Diseases caused in sheep are louping ill (wobbly gate/seizures), lamb pyaemia (cripples), tick-borne fever (fever, ill thrift, weight loss and death).

If you know ticks are a problem in your area, inspect the sheep regularly during the warmer months; if adult ticks are found on the sheep, treat with a suitable pour-on or injectable insecticide.

SHEEP DIPPING

In the twentieth century, most flocks of sheep in Britain were routinely dipped once a year in water containing organophosphates (OPs). These chemicals were developed as wartime nerve agents and then adopted for use as insecticides. The OP dips were efficient at clearing sheep of ticks, lice, sheep scab and blowfly, but by the 1990s, it was apparent the chemicals were causing long-term serious health problems in people exposed to them. While still available to licensed contractors, OPs should be avoided if possible because of the long-term risk of developing neurological conditions.

CHAPTER 18

SHEEP'S WOOL AND SHEARING

A Wensleydale lamb, a now rare longwool-producing breed from northern England.

Sheep shearing in Blaenpennal, Ceredigion, Wales in the 1930s. People gathered to help their neighbours at this important time in the farming calendar.

WOOL MARKETS

Before the invention of man-made textiles (such as polyester and nylon), the annual harvest of wool from Britain's national sheep flock was a very important income source for sheep farmers. Wool was, and still is, an important export product. British breeds of sheep were developed and improved with wool quality in mind and not just meat yield.

CARBON-POSITIVE PRODUCT

Sheep's wool from grass-fed flocks is a wonderful planet-friendly product for clothing and insulation. Sunlight, plant photosynthesis, water, soil, minerals and carbon dioxide all combine to enable the sheep on Britain's hills to produce a high-quality material simply from grazed grass. With the growing awareness of how man-made textiles contribute to microplastic pollution, landfill and fossil fuel use, sheep's wool can once again become an important and celebrated eco-friendly material.

Today in Britain, wool is marketed as a commodity by the last of the state-run marketing boards, the British Wool Marketing Board. There is a nationwide network for collecting wool – farmers deliver it to regional depots, where it is graded according to quality. The Wool Board sells most British wool at auction to manufacturers of quality wool carpets, with only the very highest quality and finest grade wools going into the clothing trade. The poorer-quality wool is also used for products like building insulation and as a fibre added to compost making.

WOOL QUALITY

If you are planning to keep sheep to produce wool for crafting activities, wool quality should be a primary consideration when selecting a breed. The British Wool Marketing Board categorises wool in up to 120 different grades or types of wool. The style of wool is generally determined by its staple (length), crimp (waviness), fineness and lustre (how much sheen it has). The workable properties and softness of the different wools is influenced by the breed of sheep. Breeds such as Southdown and Shetland produce very fine, soft wools.

Long-wool sheep, like Blueface Leicester and Teeswater, grow wool with a longer length, lustre and crimp or curl.

Some textile artists prefer breeds that produce different colours of wool, such as Jacob sheep or Black Welsh Mountain.

Wool becomes coarser as the sheep ages. Yearlings (also known as 'shearlings') and lambs offer the finest wool clip.

The wool lorry, Lampeter, Ceredigion, Wales. Farmers bring their wool to meet the lorry, which takes it to a British Wool Marketing Board sorting and grading depot.

The Teeswater breed was developed for the length and shine (lustre) of its wool.

A Jacob ewe produces a fleece of different colours. If keeping sheep for wool, assess the quality of the fleece of individual animals, as there is variation within all breeds. A sheep's fleece is both insulating and waterproofing.

Nutrition and Wool Quality

Nutrition is also a factor in determining wool quality. Sheep kept in harsher environments on rougher grazing or without adequate nutrition and minerals will produce poorer-quality wool. Sometimes mineral and nutritional deficiencies in sheep can be picked up on by looking at the quality of the fleece. Sheep with a deficient diet or affected by disease will often have dull, coarser, shorter more 'steely'-looking fleece, while the fleece of a healthy sheep that is in good body condition will have a natural 'bloom' to its appearance. The fleece will be bouncy with longer fibres and have a lustre and sheen.

WOOL USES

The fineness of wool is determined by the diameter of an individual wool fibre, measured in microns. Merino is a superfine wool and has a micron count of under twenty. Fine British wools would be in the region of twenty to thirty microns with medium wools usually thirty-one to thirty-five. The micron count influences whether the wool can be made into garments. Lower-micron-count finer wools are suitable for wearing close to the skin, while the coarser medium wools are only suitable for outerwear, unless the wool is from lambs of the breed. Breeds that fit the 'medium' versatile wool description include Llanwenog, Texel, Romney, Scotch half-bred, Lleyn and Border Leicester.

Mountain sheep breeds, like Welsh Mountain, Swaledale, Herdwick and Scottish Blackface, grow tougher but coarser wool fibres, which are ideal for hard-wearing wool carpets. Hill or mountain wool contains brittle white kemp, so is not suitable for clothing. The lowest-quality wool is used in blends for insulation. Some hill farmers have developed businesses that produce garden compost made from composted bracken and sheep's wool – the wool is an important water-holding component in the compost. A useful substance called lanolin is also recovered from sheep's wool and used in many cosmetics, skin creams and lotions. Lanolin has waterproofing and softening properties, providing a protective layer to the skin.

Whatever the end use of sheep's wool, it is the most sustainable and carbon-neutral textile material on the planet, so must have a future in a world where carbon emissions need to be curtailed.

RAW SHEEP'S WOOL

A raw fleece straight off the sheep's back is not very useful. It is smelly, greasy and usually full of leaves, twigs and bits of hay. If stored or used in this state, it will be a haven for insects and vermin.

Washing Raw Fleeces

Washing is best done outside on a nice sunny day. You will need a hosepipe, a large tub, a supply of hot water and a large drying rack to hold the wet fleece. Plenty of guides are available online demonstrating the process of home washing raw sheep's fleece. Wool is incredibly versatile, and, once the fleece is washed and dried, it can be carded and spun into yarn, felted or used as stuffing for cushions.

WHAT TO DO WITH HOME-PRODUCED FLEECES

Home washing and crafting Wash raw fleece at home for home crafting such as spinning, weaving, knitting and felting. The fleeces must be washed and processed before they can be used.

Sell the raw fleeces to a local crafting individual or group Home spinners and weavers are always on the lookout for good-quality fleeces of different colours. Values will usually be three to four times the mass market value of raw fleece. Fleeces from specialist longwool breeds, like Cotswold or Teeswater, could be worth even more to the right buyer. Try advertising to local craft groups online for customers.

Sell fleeces for commercial use The Wool Marketing Board or a processing company will also take fleeces off your hands. Values are generally low but at least the wool will be put to good use. Find out where the local collection centre is in your area.

SHEARING

Timing

In Britain, shearing is done in early summer. In the south of the country, shearing begins in early May, and the time becomes progressively later the further northwards you go – flocks in Scotland will be shorn by mid-July. The shearing date is also decided by the farm's altitude, with lowland flocks being sheared earlier than hill and mountain flocks. Ask people in your area when they normally shear the sheep.

Preparation for Shearing

Trim off the dirty wool around the back end of the sheep (known as 'dagging') a month or two before shearing to remove soiled wool from around the tail area. This will make shearing cleaner and easier and will also help prevent flystrike in early summer. Make plans for how you will market and store your wool. Wool needs to be stored in a dry shed, off the floor on a pallet or table. If supplying to the wool board, you will need to collect wool sacks from your local collection depot.

Contract shearers will often come at short notice to do small flocks between jobs. Ask sheep farmers in your area or at your local agricultural merchants for shearer contacts. Shearers normally charge per sheep, but with small flocks, expect to pay a travelling and set-up charge as well. Each season there is a shortage of skilled sheep shearers in Britain, who work extremely hard to clip all of Britain's 13 million or so breeding sheep.

Ensure you are prepared before the shearer arrives, with the sheep penned up and a clear, tidy work area ready. Shearers prefer that the sheep are fasted overnight in a barn or yard and kept off grass, which helps to keep muck away from the work area. Sheep must be dry before they are sheared, so may need to be housed the night before if showers are forecast.

Shearing a Llanwenog ram in the first week of June.

THE WOOL RISING

Shearing is much easier when the wool is rising. Sheep have a waxy, oily layer at the roots of the wool fibres, which clogs the shears and makes shearing difficult. When temperatures rise in early summer, the layer of wax rises away from the skin – often this is indicated by a yellow line on the wool fibres. When this occurs, the wool at the fibre roots becomes looser and cuts away easily. By shearing the sheep at the right time, an experienced shearer can quickly and efficiently cut away the fleece in one whole piece.

If an electrical socket is not available in the work area, ask the contractor if they can bring a generator. Also check if they bring their own handling trailer. This is used for penning a small number of sheep before dragging them onto the shearing mats. A good supply of refreshments is usually welcomed as well.

Shearing Equipment

- A large plywood board or rubber mat for shearing on
- A small catching pen, ideally with a spring-loaded or easy-close gate that shuts behind the shearer when a sheep is dragged out of the catching pen and onto the shearing board

Llanwenog sheep collected for shearing.

- A large table or clean board on which to roll the fleeces
- If using the large wool sheets or sacks from the Wool Board, a wool-packing frame is useful for holding the sack. The above

items will often be supplied by the shearing contractor, but check this beforehand
- If shearing the sheep yourself, wear moccasins, slippers or just socks to make it easier to move your feet when shearing
- Antiseptic spray for dressing any cuts left on the sheep from the shears.
- Mains-powered or 12v shearing machine plus driveshaft and handpiece. Training by a professional shearer is required in how to operate and service the equipment if you have not used it before. The British Wool Marketing Board organises courses in summer to teach people how to handle the sheep, shear the wool from the sheep and set up and service the equipment.
- Hand shears if shearing by hand; the traditional one-piece steel loop handle and 15cm (6in) blade can do the same job as a machine but takes longer

The Shearing Process

The photo sequence here shows how to shear a sheep for a right-handed person. The hand not holding the clippers is used to position the sheep and smooth out the skin for the clippers.

For efficient and quick handling, the sheep are put into a handling pen with a spring-loaded door, supplied by the shearing contractor.

The fleece is tightly rolled, with the clean side facing out.

The filled sacks, weighing 60-70kg (132-154lb) each, ready to go to the British Wool Marketing Board.

The fleeces are packed into a large sack held by a metal packing frame.

1. Standing to the side of the sheep, bend the sheep's head sharply over her right shoulder and swing the sheep towards you.

2. Roll the sheep's back end towards you with your right hand so her weight is on her left thigh and flank.

4. Shear the dirty wool from the tail and crotch.

3. Pull the sheep up by her front legs, resting her back against your legs. Tuck your left foot under her belly and shift her weight onto her left thigh. You are now ready to shear the belly, using downward strokes from right to left.

5. Shift the sheep's weight onto her right thigh. Cut an opening in the wool at the fore end of the left thigh.

6. Clip the wool from the left thigh in downward strokes, working all the way round to the backbone and towards the tail.

8. The sheep is pulled upright and sheared from the brisket up the underside of the neck to the chin. This is one of the most challenging moves to learn.

7. The back end is clear of wool. Now grip the wool of the lower neck and pull the sheep up to shear from the brisket up.

9. Turn the head upright and clip the wool from the shoulder to the cheek.

10. Clip the wool from the shoulder to the top side of the neck in long, curving strokes.

12. Gradually lay the sheep down and clip her left flank.

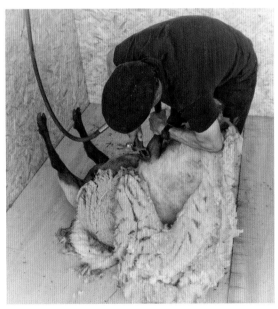

11. Pull the front leg up towards you to clear the wool at the top of it.

13. Finish the left flank with long strokes of the shears from the back of the thigh up to the neck. Keep the sheep's head pulled around your left leg to stretch out the skin on her side.

14. Step over with your right foot and shear the sheep's left flank up to the backbone.

16. Holding the sheep under the chin, pull her weight back towards your knees while shearing the wool from her right neck and shoulder.

15. Lift the sheep's head up and begin clipping down the right side of her neck. Slide her back end over with your right foot as you go.

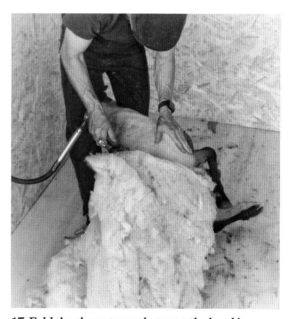

17. Fold the sheep towards you so the head is between your knees. Keep the sheep's weight on the right side of her rump and shear the wool in smooth strokes down the right shoulder and flank.

18. Clip the wool from the backbone towards the belly, slowly moving backwards and rolling the sheep towards you as you go.

20. The sheep ends up lying on her left side. Keep your toes turned in so they are under her shoulder. Finish clipping her rump and hind leg.

19. Continue clipping from the backbone across the flank and down the right thigh.

21. The fleece has come off all in one piece and there are no cuts on the sheep: a good job done!

CHAPTER 19

BASIC SHEEPDOG TRAINING

Floss, pictured here at twelve years old, was a tri-colour border collie bred from registered sheepdogs on a farm in the Brecon Beacons. Wildly effervescent as a youngster, her *raison d'être* was to chase sheep and get involved with farm work whenever she could.

There are several breeds of dog that will work sheep, but the sharp and bright border collie is the classic one that comes to mind when most people think about a sheepdog. Famed all over the world for its ability to work sheep, the border collie is a dog with a loving temperament and boundless energy.

The border collie is sensitive and intelligent and, with the right handler, has the potential to become a well-rounded, professional and much-loved working and life companion. But it's worth pausing for a note of caution here. The young border collie's potential for learning good habits and behaviours from its handler is more than matched by its potential to learn bad habits and dangerous and antisocial behaviours.

On farms and in towns, in rescue centres and people's homes, there are far too many problem collies. They exhibit behaviours such as barking, being aggressive and biting, running away, chasing traffic and failing in house training. All these behaviours are the results of the dog handler's mistakes when the dog is young, for example failing to spend enough time with a dog as it develops, lacking patience and consistency in training and an overall failure to understand the psychology and vulnerabilities of the border collie.

Before embarking on your sheepdog journey, ask yourself some honest questions. Will you have time to walk your dog three or four times a day and socialise with it? Do you have a kennel and run or safe space where the dog can go to settle down when you are busy? Do you have the patience to deal with challenging behaviour without losing your temper? Will you be able to provide the dog with consistency and routine? Are you prepared for the fact that working dogs are extremely energetic and do not often relax? There are some collie family lines that have been selected and bred for their looks and attributes as pets, and these can make better house companions. Dogs bred for working sheep can be challenging to handle in a home situation as, simply put, they want to chase and herd everything, from cars to footballs to children.

Will you be able to keep enough sheep for the dog to practise with? A sheepdog will willingly work all day every day if allowed. While this is only necessary on the largest sheep farms, any sheepdog appreciates the opportunity to 'keep it's paw in' and spend two or three ten-minute sessions a week rounding up some sheep.

Some sheepdog handlers keep a small group of hoggets (lambs over six months old), or a group of retired ewes, just for the purpose of sheepdog training. Think about whether you have enough acres to do this. To train a sheepdog, you will need at least one large field (over 0.8ha or 2 acres) that the dog can work the sheep in.

FINDING A SUITABLE DOG

In order to train a sheepdog, it must be born with an instinct to take an interest in sheep. In a good border collie, this interest will be more than just wanting to run at sheep; it will want to watch them whenever it can and maybe it will try to herd them. Even six-week old border collie puppies can show interest in sheep by staring at them, creeping towards them or running around them.

For the best chance of finding a dog with a herding instinct, try to source a puppy from a family of working dogs. This might be through a person who competes in sheepdog trials. The International Sheep Dog Society could be a good place to ask about members in your area, or find out if

there is a regional sheepdog trials organiser. A sheep farmer with well-trained dogs may also advertise puppies for sale. Always be prepared to pay a bit more when buying from someone who is knowledgeable about sheepdogs and takes the time to answer your questions.

Buying an older dog that has already been trained can be a good introduction to keeping a sheepdog.

Always visit the puppies at the premises where they were bred and see that all the dogs are well cared for and well socialised. If possible, ask to see the parent dogs working some sheep.

Be aware that many farms will sell litters of puppies simply to make money with no interest in the suitability of the dog either as a pet or working sheepdog. Perhaps surprisingly, many sheep farmers do not know how to train a sheepdog as they mostly use a quad bike for gathering sheep. Avoid puppies from farms that only use their dogs as yard dogs or for moving cattle; it is much better to source a dog from an enthusiast who is experienced at training their own sheepdogs.

A dog working on the hill needs boundless energy and determination to gather the flock over rough terrain.

Consider what type of sheepdog you want. A hill-type dog will run and herd a large flock energetically but can be hard to stop, while a trials-type dog is easier to stop when you want it to but can be less suited to moving larger flocks. If you have time, visit a sheepdog trials competition and speak to the handlers about their dogs – an invaluable way to glean information.

STARTING OFF

Give a puppy plenty of time to get to know its surroundings. The dog will need to learn some basic house rules and become socialised and introduced to different sights and sounds.

Let the dog become accustomed to its routines of exercise and feeding times by following standard guidelines on training, feeding and socialising a puppy. Don't expect too much of a young dog – it will

need time and space to have fun. Teach the dog the stand (stay in one place) and recall commands before beginning training with sheep.

HOUSING YOUR DOG

Because of their lively and mercurial temperament, working dogs are often kept in a secure run with a kennel in a barn or outhouse. A good handler will bring the dog out at regular intervals in the day to exercise, train and to be socialised. Border collies are very excitable, and some dogs will not get sufficient rest unless they are put away for a few hours each day. It is not essential to kennel a working dog, but it will need a safe, quiet secure space where it can go while you are busy.

INTRODUCING A YOUNG DOG TO SHEEP

An appropriate group of sheep is needed for training a young dog. Do not use any sheep that might be aggressive towards the dog, such as ewes with lambs, or mature rams.

Give a puppy time to socialise and get to know its surroundings before putting too many expectations of work or training on it.

Do not use ewes that are still suckling lambs for training a young sheepdog.

A good group would be between five and fifteen hoggets or older ewes. Ideally the sheep will be docile but still a bit wary, and back away from the dog. Sheep that are wildly terrified or blindly panicking and liable to jump over fences are not suitable.

Before letting the dog out of its kennel, have the sheep contained in either a circular pen made of sheep hurdles about 5–10m (16–32ft) in diameter, or in a small paddock. The idea is to introduce the dog to the sheep in a controlled situation. In a big field, the sheep will usually run away and an untrained dog will wildly run after them, completely ignoring the handler. In this type of situation, it is difficult to catch the dog and bring it under control.

The first few training sessions should only be five or ten minutes long. The purpose of them is for the dog to see the sheep and for the handler to see how the dog reacts.

Assessing Readiness for Training

In a Pen
With the sheep safely in a pen made of hurdles inside a field or paddock, walk

towards them with the dog on a lead. Once close to the sheep, let the dog off the lead. Leave the dog on the outside of the pen while you go in with the sheep. If the dog runs around the outside of the pen, excited by the sheep, it might be ready for further training; *see* Balancing, below. On the other hand, the dog may react by:

- Trying to get into the pen by squeezing underneath
- Jumping over into the pen
- Putting its paws up to the bars and yipping
- Running away in the opposite direction or showing no interest in the sheep whatsoever

It may be trying to get into the pen because it doesn't want to be separated from its master, or it wants to get at the sheep and bite them.

If any of the above behaviour occurs, it is best to leave training and to try again in a week or two. If the dog is not interested in the sheep, it probably needs more time to mature. Give the dog some more quiet kennel time in between socialising and exercising. Trying to interest a young dog in herding sheep when it is not ready can put it off for life.

Before starting training, gauge how keen and robust the young dog is mentally. Starting a puppy too early with formal sheepdog training can put it off working sheep for life. Be very patient with a young dog.

BASIC SHEEPDOG TRAINING COMMANDS

Stand = Stop by lying down or standing still
Come bye = Move around the sheep in a clockwise direction
Away = Move around the sheep in an anticlockwise direction
Walk = Walk towards the sheep
That'll do = Come away from the sheep and return to the dog handler
Heel = Walk at heel

Border collies mature intellectually and emotionally at different rates. Some dogs can be working sheep at ten weeks old while others might not be ready until they are six months.

Balancing the Sheep

As you step into the pen, the movement of the sheep may excite the dog. It runs around trying to find a way closer to the sheep. If the sheep move away from the dog, this gives it confidence and it runs to the other side of the pen to 'herd' the sheep back. A dog with a good instinct for sheep work will naturally keep the sheep between itself and its master; this is called 'balancing the sheep'.

A good sheepdog naturally wants to stop the sheep moving away from its master and keeps trying to herd them back. In this instance, if you walk around the sheep in the pen, a good dog will move to mirror the movement, keeping the sheep in between it and you. This is a good sign, for the dog is ready to progress with training.

In a Small Paddock or Field

This method is less controlled than starting with the sheep inside a circular pen. It is easier if you have a group of sheep that are used to seeing a dog, or accustomed to coming to you when you have a bucket of feed.

Young sheep not used to seeing a dog will usually run and huddle in a corner of the paddock. When walking into the paddock with the dog, analyse what the dog's reaction is to the sheep. Let the dog off the lead and see what it does.

A timid dog, lacking confidence, may just lie down and stare at the sheep. Walk around the sheep and see if the dog creeps towards them or runs around them. Hopefully the dog will run around the sheep in a herding motion. If it does, wait until it is in a position opposite you.

The dog here is 'balancing' the sheep, keeping the flock between itself and its master.

If you can hold a moment where the sheep are balanced in between you and the dog without either breaking into a run, give the command to 'stand' to the dog. This teaches it to stop on command. Make your commands crisp, clear and simple.

FURTHER TRAINING

Controlling a Keen Dog

A wild, excitable young dog may run at the sheep immediately and try to grab the wool with its teeth. If the dog chases the sheep excitedly or grips hold of them, position yourself to block the dog from the sheep.

If the dog stops for a moment, give the command to 'stand'. You may have to wait until the sheep are in a corner of the paddock or against a fence before you can position yourself between the sheep and the dog. The idea is to teach the dog to keep a distance from the sheep. If the dog moves sideways, step left or right to keep yourself in between the dog and the sheep. When the dog stops, again command it to 'stand'.

Whenever a dog becomes uncontrollable, do not shout at it; catch it calmly, put it on a lead and walk away, giving the command 'that'll do'. Remember, the dog wants to

Herding sheep relies on being able to stop the dog at the correct time. 'Stand' is the command usually given to make the dog halt and lie down. Teaching a keen young dog to 'stand' and to not move until you give the command takes time and patience.

A confident young dog will be able to run behind the sheep and 'lift' them off the fence.

impress you, it just does not know what you want!

A successful sheepdog needs to enjoy his or her work herding sheep. If a sheepdog is hit, shaken, or shouted at by its master for chasing sheep, it will not understand why it is being chastised. The purpose of training is to teach self-control, for both dog and dog handler.

Changing Direction

If the dog is choosing to run around the sheep, give a command as it changes direction. When the dog goes round the sheep in a clockwise direction, command 'come bye', and when the dog goes in an anticlockwise direction around the sheep, command 'away'. This can be taught with the dog running around the sheep in circles, or with the sheep in a corner of the paddock and the dog going back and forth from fence to fence.

Eventually the dog should gain the confidence and skill to be able to 'lift' the sheep off the fence by running between the

sheep and the fence, bringing them along as one group. If it can lift sheep off a fence line, go left and right on command and has learnt to stop when you say 'stand', the dog is ready to learn the gather.

The Gather, or Outrun and Fetch

If the dog has learnt to run round the sheep clockwise ('come bye') and anticlockwise ('away'), and is changing direction on command, the dog is ready to progress to gathering the sheep from a distance. This is also called the outrun and fetch.

Start learning the outrun by taking the dog into a small paddock or field with the sheep. As soon as the dog focuses on the sheep, make it stop by giving the command 'stand'.

On command, the dog will run in an arc to one side of the group of sheep (the outrun), and then around to the back of the flock to set them moving towards the handler (the lift or fetch). The dog will keep the sheep between itself and the handler, bringing them back down the field, flanking left or right as needed to steer them along.

Decide if you will try to send the dog 'come bye' or 'away'. You will soon find out if the dog has a natural bias/preference for going around the sheep clockwise or anticlockwise. To begin, work with the dog's natural choice.

A successful outrun and gather is achieved when the dog runs in a smooth arc around and behind the sheep and brings them towards the handler.

Starting the Outrun

Make sure the dog stays in a 'stand' position, then walk towards the sheep. If the dog follows or runs off, reset it in the original position and repeat the command 'stand'. This can take a lot of patience and may require more than one session before the dog learns to stay still.

Stand a few metres away from the dog with your back to the sheep. With eyes still on the dog, turn your body to face the direction you wish the dog to take around the sheep and give the command 'come bye' or 'away'. To begin with, the outrun should only be approximately 30m (100ft) long. As the dog progresses, the length of the outrun can be increased. The best sheepdogs will run and gather sheep that are a few hundred metres away, but this takes time and regular practice to learn.

Common Problems with Learning the Outrun

The dog will not set off on its run and creeps or stays still The dog needs to return to an earlier stage of training to build confidence and learn that chasing sheep can be fun, or it needs a break for a few weeks to further mature.

The dog runs straight at the sheep and scatters them or singles one out Take the dog back to work in a small pen or paddock. If it dives in to scatter the sheep, try to block it and command 'stand'. Make the dog run in a wider arc around the sheep by adjusting where you stand at the start of the outrun – a few metres in front of the dog and a few metres to the side in the direction you wish the dog to run. This encourages the dog to run around you first and then the sheep and teaches a better shape to the outrun.

The dog does a nice outrun but then runs round and round the sheep A dog 'playing' with the sheep instead of fetching them and bringing them to the handler is usually behaviour it will grow out of. If this problem persists, give the dog a break from sheepdog training. Take it back to basics and teach it to 'stand' on command again. Return to working on a shorter outrun with the dog and sheep at closer quarters to regain control, and teach 'stand', 'come bye' and 'away' again.

The Drive

When the dog drives the sheep, it walks them away from the handler. This is a useful skill when you need to move the sheep through a gateway or into a catching pen.

Teaching the drive is easier if you have a straight fence line on one side of the sheep so they can only turn to one side.

Start with the sheep standing next to the fence line. Have the dog lying down at your feet in the 'stand' position. Start moving towards the sheep and command 'walk' to the dog. Ideally, the sheep will start walking away up the fence line away from the dog. This should be done at a steady pace. If the sheep try to break free on the unfenced side, the dog should flank them (block their progress) with a short 'come bye' or 'away' movement. Asking the dog to 'stand' and 'walk' at the correct moments will control the speed and direction of the sheep.

Learning to drive the sheep away from the shepherd is a more advanced skill for the dog because it overrides a sheepdog's strong instinct to always bring the sheep back to its master.

See how far you can get along the fence before the dog excitedly darts right around the sheep and brings them back to you. The skill of driving sheep relies on having an effective 'stand' and 'walk' command over the dog.

The Pen

The final stage of gathering the sheep is to secure them in a pen. As with the drive, good control needs to be achieved over the dog before learning to pen the sheep. As the handler stands at the end of the gate ready to shut the sheep in the pen, the dog will need to use all the skills it has learnt to guide the sheep in: flanking left and right, stopping on

Using the 'come bye' and 'away' commands to keep the flock together, tucking in stray individuals on the flanks. A dog is fast enough to prevent sheep breaking away at the sides and keep the flock moving in the desired direction.

command, pushing the sheep on with a walk. Penning the sheep needs to be done in a slow and controlled manner and is not a skill a young dog will achieve quickly unless it is of high calibre.

SHEEPDOG TRAINING TIPS

- Keep training sessions short but regular.
- Deliver training commands clearly and crisply.
- Keep commands clear and simple.
- Keep your gaze focused on the dog when it is working.
- Patience is essential when training a young sheepdog.
- Once the dog is trained to follow voice commands, take the time to train it to a whistle.
- Keep training sessions short (five to fifteen minutes) but regular.
- Beware of duplicating commands. When I took my young collie on a sheepdog-training course, the first observation the teacher made was that I was using variously 'lie down' 'stay' 'sit' or 'stand', where just one of these would do. Too many commands will only confuse the dog.

FINDING A GOOD TRAINER

If you are beginning as a complete novice with a young, untrained sheepdog and your own sheep, a few sessions with a sheepdog trainer could be invaluable in starting off. Trainers will have different methods and ways of keeping and training dogs. Ask them to explain how they would start training with a young dog and you will be able to see if they share a similar approach to how you keep your dog.

The largest part of sheepdog training is teaching the dog to stop (stand) when you need it to. A lot of patience is required to teach a young dog to stand while it is in the thrill of the chase.

If the dog seems wilfully disobedient and fails to follow a command that it has learnt before, do not chastise it, end the training session immediately, and try again another day.

If the dog is failing to pick up on what the handler is trying to teach it, the dog is either not mature enough to learn the lesson at that time or the handler is trying to teach it the wrong way.

The International Sheepdog Society is a good source for training materials and contacts. Watch sheepdog trials on TV to get an idea of how the pros do it.

ACKNOWLEDGEMENTS

This book couldn't have happened without my father, Jim Cockburn, providing the land and tools for me to embark on this farming journey. Thank you for founding a family farm and letting me do as I please.

Added to that, thanks for the invaluable support of family and friends who unwittingly discovered that a farm is not just run by a farmer: it requires a whole team to make it work. In preparation of this book, huge thanks to Leah Cockburn for photography, and to Leah, Lauren and Annie Cockburn for help with lambing and looking after the flock. Thanks to the grannies, Barbara and Linda, for your family support, and to Peter 'Jules' Thewless who, despite still needing to be convinced about the merits of sheep, has always stepped in to help anyway.

Thank you to our knowledgeable sheep-farming neighbours, Philip and Hilary Thornely, and John Green for loaning me the first ram. Finally, enormous gratitude to Eleanor Bryant for her brains and for putting up with it (us) all.

INDEX